How often and where and why have there been scenes like this on Earth durig the past three billion years?

Part of the Antarctic Ice Sheet in the Sentinel Range, looking southward from Whiteout Nunatak. January, 1967.

Geological Society of America
Memoir 192

Pre-Mesozoic Ice Ages:
Their Bearing on Understanding the Climate System

by

John C. Crowell
Department of Geological Sciences
and
Institute for Crustal Studies
University of California
Santa Barbara, California 93106, USA

1999

Published by The Geological Society of America, Inc.
3300 Penrose Place, P.O. Box 9140, Boulder, Colorado 80301

Printed in U.S.A.

GSA Books Science Editor Abhijit Basu

Library of Congress Cataloging-in-Publication Data

Crowell, John C.
    Pre-Mesozoic ice ages : their bearing on understanding the climate
system / by John C. Crowell.
      p. cm. -- (Memoir / Geological Society of America : 192)
    Includes bibliographical references.
    ISBN 0-8137-1192-4
    1. Glacial climates. 2. Geodynamics. I. Title. II. Series:
Memoir (Geological Society of America) : 192.
QC884.C75 1999   99-35888
551.6'09'01--dc21   CIP

Frontispiece: How often and where and why have there been scenes like this on Earth during the past 3 b.y.? Part of the Antarctic Ice Sheet in the Sentinel Range, looking southward from Whiteout Nunatak. January 1967.

10 9 8 7 6 5 4 3 2 1

Contents

Preface

During the past half-century much has been learned about the Earth throughout the long duration of geologic time, and I have enjoyed being involved in finding out about its unfolding history. Research during and following World War II led to many advances, including a better understanding of sedimentation processes within the seas and oceans, and the realization that sand and coarse gravels were carried into deep waters by turbidity currents and by downslope sliding. By the early 1960s I was quite sure that many reported ancient glacial tillites described from far-flung stratigraphic sections over the Earth were probably not glacial in origin, but were debris flows or the result of downslope sliding. I became a disbeliever in some ancient glaciations and launched studies of late Paleozoic stratal sequences in the Southern Hemisphere, believing that many would prove to be nonglacial in origin. By the late 1960s, however, I had examined critical sections on Gondwanan continents and became convinced that huge glaciers had indeed scoured across the supercontinent of Gondwana as other geologists had concluded before me. Paleomagnetic studies and better dating were accumulating and showed that during the Paleozoic era, the supercontinent had moved across the South Pole and that the movement of cratonic blocks could account for much of climate change, and especially the growth and decay of continental glaciers through time. In the 1980s I changed my focus from the clear record of the Paleozoic to the record of all pre-Mesozoic times of iciness. How fortunate have I been to peer at telltale strata on all continents in the company of friendly and stimulating regional experts!

In the 1960s plate tectonics came along and I was most receptive to the concept because much of my work along the San Andreas fault system had shown that large horizontal movements were probable. Geoscientists began to recognize that crustal blocks were moving about in an understandable scheme. These processes, involving sea-floor spreading, sliding of the crust down subduction zones, and the lateral movement of plates along transform boundaries, were deemed at first to be characteristic of crustal activity during late Phanerozoic time only. The record preserved on the sea floor in the form of magnetic-anomaly patterns, could be followed back in time only into the early Mesozoic. The record of older crustal activities is preserved only in bits and slices of strata caught up and incorporated into continental plates. Geoscientists now advocate a similar crustal mobility on back to the beginning of the stratigraphic record, more than three b.y. ago.

During the past several decades dramatic advances of knowledge of space history and processes have come about, and satellite imagery and computers have been applied to understanding the Earth. Studies in critical areas over the world by many geoscientists using many approaches now lead us to a picture of how continents with mountains and basins and plains upon them have moved about throughout geologic time. Historical geologists conclude that tectonic processes of the past were not too different from those operating today. These complexly interwoven processes have gone on for more than 3 b.y., but with interruptions and changes in pace and style from time to time. Bolides have impacted the Earth and density stratification of ocean waters has overturned. The guiding concept of gradual and continuous uniformitarianism is now replaced by a concept of interrupted uniformitarianism. Biogeochemical changes have affected the seas and atmospheres and the quantities of greenhouse gases. Ocean currents have been rearranged owing to changes in the thermohaline "conveyor belt" and as gateways between cratonic pieces have opened and closed. Earth orbital perturbations and other extraterrestrial influences have taken place. Paleontological and geochemical research into Proterozoic time and before has flourished during the past half-century and I have been stimulated in reading about the discoveries. Carbonates capping Neoproterozoic tillites pose perplexing geochemical and climatological questions. Nonetheless, the surface history of our planet, including its climate, has only briefly departed from a comfortable mode. Throughout all of this, life of many forms has continued to prosper and evolve. The atmosphere-ocean-land system has always functioned within limited bounds.

Here, in this memoir, I review briefly what geologists have learned concerning the cool part of the climate record, the part recording ancient ice ages. We have documentation of cool episodes when tongues of ice from glaciers reached the sea interspersed with longer intervals of relative warmth. Data are still coming in so this is only a progress report and future research will provide elucidation. The history of climate extremes as revealed by times of widespread iciness through 3 b.y. of geologic time purports to tell us much concerning how our Earth system works.

My personal involvement over the past 60 years in this growth in understanding the Earth's surface history has made this a stimulating time to be a historical geologist. I have fortunately lived during what may go down in human history as the "science century." I was first educated as a field geologist, and had nearly two year's professional field experience mapping in complex California terranes when World War II came along. I was then reeducated as an oceanographic meteorologist for wartime service, and was involved in forecasting sea and swell and surf for the Allied Normandy Invasion of 1944 and other landing sites over the globe. These experiences provided me with an appreciation of the complicated ways the air and ocean system works today. After the war I returned to geologic investigations on land, and studied tectonics and sedimentation patterns along faults of the southern reaches of the wide San Andreas fault system in California. Basins along this splintered system opened to receive thick sedimentary piles eroded from nearby uplifted margins. Turbidity currents carried sand down submarine canyons onto basin floors. Steep slopes favored sliding of sedimentary debris into the depths, and some of these deposits resembled glacial deposits at first glance. As crustal slices moved for great horizontal distances within this mobile transform belt, deep-water strata were squeezed and raised and then exposed so that geologists could crawl over outcrops and make interpretations. I became interested in the questions: How do we distinguish glacial deposits from debris-flow and other downslope deposits and how do we recognize them with confidence?

I now appreciate that tectonics, continental movements and rearrangements, sea-level ups and downs, biogeochemical changes including fluctuations in greenhouse gases, bolide impacts, Earth orbital cycles, insolation variations, water overturns in deep ocean basins, and other processes contribute to a complicated surface and climate history. Geoscientists have just begun to appreciate how these many processes interact and to describe this long history. Unfortunately, much of the rocky record has been eroded away and is no longer present to reveal this history. Future understanding of our planet's long life will come from many approaches, especially better isotopic and biogeochemical information and dating and paleomagnetic studies that will allow better paleogeographic reconstructions. The pre-Mesozoic glacial history of our planet is here reviewed as now partially understood.

During this same past half-century science has harvested the fruits of reductionist approaches, and paleoclimatology has hugely profited. For example, isotopic analyses reveal geochemical traces of ancient temperatures in fossil shells. In the future no doubt additional indications of ancient climates through geochemistry and other approaches will come along. We now need to learn more about the time distribution of greenhouse gases, especially water vapor, carbon dioxide, and methane. The role of changing albedos through time and the frequency and effects of bolide impacts present challenges. The door to understanding climate through the long reaches of geologic time is just opening. We now know with confidence that the climate system is exceedingly complex and affected by many feedback loops, but the way the loops operate and how they have operated through time is just now being learned. Appropriately, geoscientists are now focusing on complicated systems where many processes and factors interact. Concepts involving stasis, plate-tectonic rate changes and rearrangements, modifications of equilibrium through time, interrupted uniformitarianism, punctuated equilibrium, chance, randomness, complexity, earthly homeostasis, geophysiology, and self-organized criticality of such systems are drawing attention. During most of my career I have been an ardent adherent to the uniformitarian views of Hutton and Lyell, but now I recognize that infrequent events such as bolide impacts and huge volcanic outpourings demonstrate interruptions in the slow and steady progress of earthly processes. A main challenge now facing Earth historians is to understand how these many processes interact and what telltale products they produce.

John C. Crowell
Santa Barbara, California

Geological Society of America
Memoir 192
1999

Pre-Mesozoic Ice Ages:
Their Bearing on Understanding the Climate System

John C. Crowell

Institute for Crustal Studies and Department of Geological Sciences, University of California,
Santa Barbara, California 93106, United States

The record of ice ages throughout 3 b.y. of Earth history reveals that climate is largely the result of tectonobiogeochemical interplays within the complex crustal, oceanic, and atmospheric system. It is also moderated by orbital variations, and occasionally interrupted by bolide impacts.

ABSTRACT

Pre-Mesozoic glaciations on Earth are documented by identifiable glacial debris preserved within the stratal record, by rare glacial pavements and landforms, and by geochemical proxies indicating times of climate coolness. Although much of the record of ancient cool intervals has been worn away following uplift and erosion, documentation remains in sequences laid down in basins where strata have been preserved through long periods of geologic time, and then later uplifted and exposed for our examination.

Crustal tectonic mobility, similar to plate-tectonic activity today, is deemed to have operated as far back as the icy record has been identified, to nearly 3 b.y. ago. Glaciers are viewed as always having occupied some continental areas, usually mountains, throughout this long duration. From time to time glaciers have expanded to reach the sea and have left a stratal mark showing times of coolness that interrupt irregularly times of relative warmth.

Studies of the Late Cenozoic Ice Age, which began in late Eocene time, disclose complex interrelations between continental positioning and topography, oceanic bathymetry and location of gateways between land masses, organizations of both air and ocean currents, and concurrent effects of evolving life and its biogeochemical products. Research on the Cenozoic record is revealing processes and products to guide interpretations of cool climate intervals during the past nearly 3 b.y. of Earth history.

Iciness on Earth is reviewed here by working back into time from the present because the record becomes dimmer and is less securely dated the older it becomes. The Cenozoic Ice Age, within which we are now living, began at ca. 43 Ma and was preceded by a warmer interval of ca. 70 m.y. back into mid-Cretaceous time. The next older Mesozoic icy intervals are Early Cretaceous (ca. 105–140 Ma) and Jurassic (ca. 160–175 Ma and ca. 188–195 Ma) and were preceded by relative warmth for ca. 60–5 m.y. before the end of the Permian period. The Late Paleozoic Ice Age waxed and waned over some 82 m.y. between ca. 256 Ma and 338 Ma. It was preceded by a warmer interval lasting 15 m.y. Iciness expanded during early Carboniferous and Late Devonian times for ca. 15 m.y. between 353 Ma and 363 Ma. Relative warmth prevailed for the next older interval for nearly 61 m.y. and followed the Ordovician-Silurian

Crowell, J. C., 1999, Pre-Mesozoic Ice Ages: Their Bearing on Understanding the Climate System: Boulder, Colorado, Geological Society of America Memoir 192.

strong and short ice age of some 16 m.y. between ca. 429 Ma and 445 Ma. For the next 66 m.y., no ice is recorded.

During the 430 m.y. of the Late Proterozoic, and lasting into the Cambrian, there were three or four ice ages (ca. 520–950 Ma). At some localities glaciation occurred at low latitudes. Previously, for about 1250 m.y. from ca. 950 Ma back to nearly 2200 Ma, no record of glaciation has yet been confidently documented, posing a paleoclimatological problem. Earlier glaciation during the Paleoproterozoic (ca. 2200–2400 Ma) is demonstrated in three or four far-distant regions. The oldest glaciation so far recognized on Earth (ca. 2914–2990 Ma) is recorded in strata of the late Middle Archean, nearly 3 b.y. ago.

These ice ages occur with irregular repetitions rather than at time-cyclic intervals. Their causes are viewed as rooted in nonperiodic thermodynamic overturns deep within the Earth that result in a mobile crust and changing fluid fluxes to the air and oceans, including carbon dioxide. Climate, including times of unusual iciness, has responded to tectonic plate rearrangements, continental drift and fragmentation, sea-level changes, and especially biogeochemical changes resulting from the winding pathways of life's evolution. Climate has also been modulated by orbital variations and perhaps by variations in the tip of the spin axis and other extraterrestrial events. It has been drastically affected temporarily by bolide impacts from time to time. Climate upon the Earth over the past 3 b.y., however, is primarily the result of changing tectonobiogeochemical activities rooted within the complex earth-air-ocean system itself.

INTRODUCTION

What seest thou . . .
In the dark backward and abysm of time?
Shakespeare
The Tempest, Act. I, Scene 2

Our agenda [as scientists] is simply to figure
out how things work.
Bak, 1996, p. 36

Glaciers upon the Earth's continents have expanded and then retreated from time to time at least since mid-Archean time ca. 3 b.y. ago (*Frontispiece,* Fig. 1). Here I examine briefly the record of ancient ice ages, especially those older than Mesozoic, and appraise environments and processes controlling their waxings and wanings. An understanding of ice ages throughout this long span of geologic time, preserved primarily in the sweep of recognizable and distinctive sedimentary facies, portends to reveal much concerning how the climate system works. The glacial record is valuable in documenting climate variations toward cool extremes and can provide insights into the many interwoven influences on climate change.

The first probable glaciation, so far recognized, occurred ca. 3 b.y. ago. Since then there have been several cool and glacial intervals interspersed between longer, warmer intervals. At present we are living in the midst of a long ice age, the Late Cenozoic Ice Age, but one characterized by cooler ice stages separated by warmer interglacial stages and within an ice age that has not yet ended. Our modern glacial times began during the Eocene epoch at ca. 43 Ma when the Earth's climate gradually entered a cooling mode after a long warmer interval (Frakes, 1979; Frakes et al., 1992; Eyles, 1993; Eyles and Young, 1994). (The time scales used within this treatise are adjusted between those of Palmer, 1983, Cowie and Bassett, 1989, and Harland et al., 1990.) The purpose of this review is to describe briefly the long record of ancient glaciations on Earth, with emphasis upon those occurring before Mesozoic time, and what this record reveals concerning the terrestrial and extraterrestrial factors causing or controlling climate change. I hope the review will add understanding of the way the climate system has evolved and operates and provide insight on its complexity.

I conclude that since at least 3 Ga there have always been glaciers somewhere on Earth wherever there were sites in appropriate positions to receive snowfall lasting from season to season that then became consolidated into ice. Mountain ranges bordering coastlines were especially likely to harbor glaciers. Ever since cratons developed with mountains and highlands upon them, and began moving about within a mobile crust upon an Earth with a watery ocean and a cool atmosphere, at some places local temperature and moisture regimes permitted ice to last and glaciers to grow. No longer should geoscientists consider glaciation as unusual; they should embrace iciness someplace on Earth as normal.

Over the past several decades geologists have documented crustal mobility and the evolution of life forms during this long time span. Our concept of tectonic mobility now has evolved from inferred continental drift into modern plate tectonics. Life evolved during this same long interval, bringing about influential biogeochemical changes in the air and oceans that are now documented well back into Archean time. Climate has both influenced and responded to these changes. Only slowly have geologists come to realize with fair confidence that glaciation is probably not an unusual event in Earth history although the preservation of a convincing direct record is indeed rare. Glaciers probably have always been lodged among mountains whenever mountains have occupied appropriate climatological positions, even in equatorial regions like those today in Africa (Kilimanjaro) or upon the high ranges of New Guinea. Only during several cool intervals in Earth's climate, however, has ice expanded to cover huge areas and form wide lobes at sea level. These lobes have left a record in strata laid down in deep, perhaps distant basins.

The tectonic regime prevailing both locally and regionally is a primary control of climate, and plate-tectonic understanding provides a useful framework for visualizing where sites occur to harbor glaciers, as emphasized by Eyles (1993). Plate-tectonic classification helps in recognizing where deep basins may preserve thick sedimentary piles and protect the glacial record within the crust for long periods from destruction resulting from uplift and erosion. The evolution of plate-tectonic concepts has gone hand in hand with increased understanding of glacial sedimentation and glaciation through time. The ancient surficial record is continually worn away by erosion to make younger sediment so

that this record is inherently imperfect. More recent geologic events are more likely to be preserved although tell tale deposits laid down in deep basins may not yet have been tectonically uplifted and exposed so that geoscientists can study them. As we look back into the ancient climate record, because glaciers occurred mainly on high ground, our sampling of the record becomes poorer and poorer. This chain of reasoning enforces the view that glaciation in the past may have been more prevalent than the direct record reveals. Fortunately, geochemical proxies of climate and environmental interpretations from fossil fauna and flora help to fill in gaps. Future interpretations of ancient climates will therefore come more from geochemical studies, especially isotopic excursions, and from paleontological investigations aimed at environmental reconstructions than from direct evidence of iciness within the rock record.

Ice ages and other major climate changes are primarily the consequence of a combination of variable causes rooted in changes in terrestrial tectonobiogeochemical regimes and in extraterrestrial

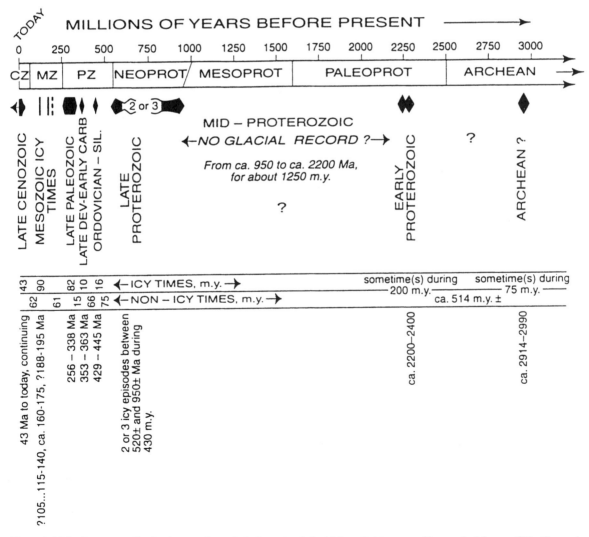

Figure 1. Major ice ages on Earth when continental glaciers extended widely so that tongues of ice reached the sea. CZ = Cenozoic; MZ = Mesozoic; PZ = Paleozoic; NEOPROT = Neoproterozoic; MESOPROT = Mesoproterozoic; PALEOPROT = Paleoproterozoic.

events and modulations. No ice age is the result of a single cause, but the combination of several, including complex feedback mechanisms. The geologic record back into the Archean shows that climate has on average ranged between limits that were neither too hot nor too cold so that life has been able to evolve and flourish. These limits have been transgressed only now and then, and only temporarily, such as at times of bolide impact. Disruptive events rooted in the mechanics of the earth-atmosphere-ocean system may also have occurred, such as when Late Ordovician bottom waters from the deep ocean probably overturned and suddenly flooded continental shelves (Brenchley et al., 1994). Rapid climatic and biotic changes occurred as a consequence. Climate is also affected and modulated by extraterrestrial influences such as variations in the Earth-sun orbital arrangements and the quantity of solar radiation reaching the Earth's surface owing to changes in the sun's output, or perhaps inhibited by cosmic dust. Despite occasional disruptive events, including times of voluminous volcanism, climate has always swung back to within limits amenable to the continuing evolution of life and the functioning of the earth-ocean-atmospheric system in a way comparable to that operating today.

The climate system is complex today and only incompletely understood. Broecker (1997, p. 6) writes: "We have greatly underestimated the complexity of [the climate] system. The importance of obscure phenomena, ranging from those that control the size of raindrops to those that control the amount of water pouring into the deep sea ... makes reliable modeling very difficult, if not impossible." But how much more complicated are all the interwoven interplays of the past 3 b.y.! Today the climate is controlled by variations in the flow of air and ocean currents upon an Earth with a known continental arrangement, known land topography, and known ocean and sea bathymetry. Studies of the Recent and Pleistocene records, when the topography and bathymetry were but little different from the present, are now disclosing how this complex and interacting land-sea-air system operates and how it is influenced by variations in orbital arrangements and terrestrial events such as volcanism and intricate feedbacks.

In examining the long record of the geologic past, comprehension of influencing factors is considerably more difficult. Continents have moved about into different arrangements and latitudes. Gateways between continental blocks have opened and closed to guide ocean currents and rearranged heat distribution in the ocean-air system. Highlands and mountain ranges upon continents have had differing positions. They have risen and have been eroded away. Ice ages, as parts of this complex scheme, are the result of the interplay of many processes and events that will never be documented satisfactorily mainly because the record is piecemeal and older parts of the stratal record have been destroyed by erosion, or are deeply buried and inaccessible. Despite these many difficulties, a look at the long record of climate change, with special attention to the waxing and waning of major recorded glaciations, will reveal much concerning the workings of the climate system. This understanding can help even in appraising the long-term influences of minor perturbations brought about by the ephemeral activities of

humankind, such as the addition of carbon dioxide to the atmosphere from burning fossil fuels.

Significance of ancient glaciations in the history of geological research and in understanding the climate system

The recognition that glaciers were more extensive in the geologic past than at present and took place in unexpected regions has played a special role in the evolution of concepts basic to the geosciences. Observations and discussions bearing on Pleistocene glaciations beginning even before the nineteenth century provided criteria useful in investigating earlier ice ages. In the late eighteenth century, Swiss naturalists recognized that polished and striated surfaces, and the position of abandoned moraines, required that alpine glaciers had previously reached far down their present valleys (Flint, 1965, 1971; Chorlton, 1982; Totten and White, 1985; Calkin, 1995). Large boulders, now found many tens of kilometers beyond the snouts of glaciers today were first called *erratics* by Saussure in 1779 (see Flint, 1971; Hsü, 1995). Much discussion ensued among Swiss and European scientists during the early part of the nineteenth century and led to suggestions that erratic boulders strewn about in northern Europe came from glaciers now melted away, or from floating icebergs. This interpretation was recognized in Norway in 1824 for deposits well beyond the extent of present glaciers and in Germany in 1832. In 1834 de Charpentier in Switzerland read a paper before the Helvetic Society strongly reenforcing the concept of widespread ancient glaciation, but his work was not published until 1841. Louis Agassiz in the meantime addressed the Swiss group in 1837 (published in 1840), and advocated "a great ice period," and became a vigorous advocate of the ice age in both Europe and North America.

In Britain, large boulders spread here and there over the countryside had long been credited to icebergs that "drifted" across a sea or lake covering the region, dumping their debris as they melted—hence, the term *drift* for their deposits. James Hutton, John Playfair, Charles Lyell, and Charles Darwin all commented on the origin of the material (Flint, 1971). It was not until the 1860s, however, that several geologists showed that glaciers themselves moved across much of the British Isles rather than icebergs drifting in the sea only, and the iceberg-drift explanation for most of the material was largely discarded. The concept of a time of coolness in the recent geologic past and the name the *Great Ice Age* were established. Much of this early discussion involved appraisal of evidence for the biblical deluge (Noah's flood).

In North America, deposits in the Ohio Valley now known to be glacial were first described and illustrated by Volney (1803, 1804). Drake in 1815 (and 1825) advocated an iceberg-drift origin for the erratics and associated debris, but, as in Britain, it was not until the 1860s and 1870s that the advance and retreat of true glacial lobes were established. Whittelsey in 1848 published the first maps showing the distribution of glacial features (Totten and White, 1985).

Evidence for *pre-Mesozoic glaciation* was first described in 1856 for Permian strata in India where coal beds, interbedded with tillitic debris, lay upon striated pavements (Blanford et al., 1856) although there had been speculations before this on the likelihood of pre-Pleistocene glaciations (Harland and Herrod, 1975). This discovery in India came while widespread debate was still underway in northern Europe and North America concerning criteria for recognizing ancient glacial deposits of Pleistocene age and so focused the attention of geologists over the world on such ancient field associations (Ramsay, 1855, 1880). Not long thereafter Selwyn (1859) in Australia, Sutherland (1870, 1871) in southern Africa, and Derby (1888) in South America documented similar evidence for Permian continental glaciations. These discoveries were startling because the evidence was found in regions far from the present poles and in now tropical and subtropical regions. They documented quite convincingly that in the remote past, during Carboniferous and Permian times, glaciers were widespread. As geological observations accumulated during the following half-century, the concepts of either very marked climate change over the globe or of marked mobility of the continents to account for these occurrences stimulated geologists' imaginations. Even more ancient glaciation was recognized on Islay, Scotland, by J. Thomson (1871, 1877), but its Proterozoic age was not yet established (Harland et al., 1993). The Proterozoic "Reusch's moraine" with its underlying striated pavement, exposed along the shores of Varangerfjord, northern Norway, was first described by Reusch in 1891, and since then its origin, referred to below, has been much debated.

Ancient glaciations even entered into discussions near the end of the nineteenth century concerning the age of the Earth (Hubbert, 1967). Lord Kelvin (W. Thomson, 1862) calculated and vigorously defended the concept that the Earth could be no older than 20–40 m.y., basing his calculations on the premise that the Earth had been cooling steadily since its hot origin (Hubbert, 1967). Scientists at that time were unaware of heat contributions from radioactive decay. Chamberlin (1899), among others, pointed out that strong glaciation in late Paleozoic time, and again in Pleistocene time, did not indicate monotonic cooling. Geologists at that time maintained that the base of the Cambrian system was about 600 Ma. With modern methods depending on radioactive isotopes collected from carefully mapped sequences, the beginning of the Cambrian period is now placed at 544 Ma (Bowring et al., 1993). This age is remarkably close for an estimate made 100 years ago!

Wegener, beginning in 1912 when he first put forward his continental drift hypothesis, drew heavily on reconstructions of ancient climate and glacial belts. His reconstructions were based on the concept that climate belts of the past were similar to those of today, and that continental blocks had moved about beneath the climate belts on an Earth with a mobile crust. He favored this concept rather than the one that regional climates had changed drastically on an Earth with fixed continents. Du Toit (1921) traced the record of Carboniferous glaciation on Gondwana. Because it

is largely found at low latitudes, he concluded that the glaciation was *"nowhere equatorial, but almost wholly extra-tropical"* and *"the ... portions of Gondwanaland actually disrupted and* [were] *forcibly torn apart from one another subsequently to the Triassic"* (his italics, 1921, p. 219). In the last edition of Wegener's book *Die Entstehung der Kontinente und Ozeane* (Wegener, 1929, Biram translation) Wegener draws heavily on Du Toit's (and others') work on ancient glaciations, and includes two maps (Wegener, 1929, figs. 35 and 36) showing the reconstruction of continents in both the Carboniferous and Permian with ice centers in high latitudes. Wegener writes: "the Permo-Carboniferous traces of glaciation, [and] also the total climatic evidence of that period falls into place with the application of the [continental] drift theory and forms a climatic system which corresponds completely to that of today, provided the South Pole is displaced to southern Africa" (Biram translation, 1929, p. 144–145). Du Toit's 1937 book, *Our wandering continents,* puts forth the case for continental fragmentation and continental drift even more convincingly, employing several lines of evidence and argument, including maps reconstructing the distribution of Permian-Carboniferous glaciations. The significance of these glaciations therefore occupies a special place in the evolution of concepts of continental drift into modern plate tectonics.

In the 1920s and 1930s much skepticism concerning Wegener's hypothesis of continental drift, especially in North America, hinged on interpretations of parts of the stratal record purported to support glaciation, but at localities deemed incompatible with the hypothesis (e.g., Van der Gracht, ed., 1928). Many sequences consisting of dispersed pebbles embedded within a muddy matrix were identified as glacially derived till but did not fit in with the continental drift concept and so the idea was discarded. The recognition that such beds, usually termed "tillites," could be formed in many ways, had not yet been elucidated. Even Wegener (1929, p. 124), however, discusses "pseudo-glacial" conglomerates formed by "ordinary" depositional processes. The intellectual tone in regard to recognizing evidences of ancient glaciation seems to have been characterized by Wegener's statement (1929, p. 124): "Generally, it is usual to regard the rock as certainly glacial only if one has been able still to detect the polished surface of the outcrop under the boulder clay of the ground moraine." For several decades thereafter, the view prevailed that only if such criteria could be firmly documented, was a glacial episode confirmed. Long after Wegener's time, prudent geological conservatism dictated that a glacial explanation was acceptable only with such indisputable evidence at hand. I now maintain that such a view is unwarranted. As outlined below, investigators now have many more criteria to document an ancient ice age. The documentation of glacial events comes in addition from the identification of widespread glacial facies and geochemical proxies showing coolness in dated stratal sequences without an underlying glacial pavement. Nonetheless, all major glacial intervals of the Phanerozoic and Proterozoic are documented someplace on Earth by glacially polished and striated pavements below tillites.

In the 1950s new data and concepts and interpretations as the result of investigations of sedimentation processes, including turbidity currents and processes of deep-water deposition, showed that some stratal units, previously identified as "tillites," were better identified as "tilloids," or units that looked glacial in origin but were not. Studies began of stratal sequences, some of which were previously considered glacial in origin, with the purpose of understanding their environment and manner of deposition (Crowell, 1957; Dott, 1961, 1963). Several different environments were documented in which downslope sliding resulted in layers of mixed pebbles and mud, which end up as tilloids. Criteria for recognizing ancient glaciations were elucidated (e.g., Hambrey and Harland, 1981). Intensive studies of modern environments, especially in marine regions near glaciers in Arctic and Antarctic areas, have taught us much concerning what to look for in the ancient record (e.g., Anderson, 1983; Molnia, 1983a; Dowdeswell and Scourse, 1990; Anderson and Ashley, 1991; Brodzikowski and van Loon, 1991; Huggett, 1991; Miller, 1996; Menzies, 1996). Studies of cores from the sea floor, including the deep ocean, added criteria for recognizing cold climate and glaciation far from the glaciers themselves (e.g., Warme et al., 1981). Geochemical data now also provide information on worldwide influences of cold climates (e.g., Gregor et al., 1988). It is therefore appropriate to take stock of what we now know concerning the history of ancient ice ages and their causes and controls.

Introduction to the record of glaciation throughout geologic time

Several distinct episodes of widespread glaciation on Earth are now recognized, separated by warmer intervals when glaciers retreated but almost certainly did not disappear from highlands or mountain ranges completely. Because paleogeographic reconstructions become increasingly less accurate as we go back into geologic time, the ice ages are described briefly in turn below working from the Cenozoic into the Archean. Moreover, for ancient ice ages evidence to differentiate between mountain glaciers, piedmont glaciers, ice caps, or widespread ice sheets is usually lacking.

The *Late Cenozoic Ice Age* is well reviewed elsewhere and so is not dealt with here in detail although studies of both modern and Cenozoic glaciations lead investigators in what to look for in the ancient geological record (e.g., Frakes, 1986; Ruddiman and Wright, 1987; Crowley and North, 1991; Huggett, 1991; Frakes et al., 1992; Eyles, 1993; Hay, 1996; Calkin, 1995). After warm times in the Late Cretaceous, cooling set in and average Earth temperatures declined irregularly to the late Eocene (ca. 43 Ma). This trend is shown by ice-rafted debris carried from Antarctica and recovered from ocean-floor cores. Cold surface waters flowed into the ocean from glaciers rich in the light isotope of oxygen (Browning et al., 1996). By the end of the Eocene Antarctica harbored ice caps that reached the sea. In the far Northern Hemisphere, ice had grown in the Greenland and Scandinavian regions by 7 Ma as shown by till, glaciomarine diamictites, and ice-rafted debris in marine cores east of Greenland (40° W long,

63.5° N lat) (Larsen et al., 1994). In North America and Eurasia a complicated waxing and waning of ice caps and sheets took place and complex advances and retreats of ice fronts accompanied rises and falls of sea level (Wright, 1989). As the Alpine-Himalayan chain of mountains rose, beginning at ca. 40 Ma, the flow of air was affected, including a strengthening of the Indian monsoon (Raymo et al., 1988; Ruddiman and Raymo, 1988; Molnar and England, 1990; Raymo, 1991; Raymo and Ruddiman, 1992; Burbank et al., 1993; Kutzbach et al., 1993; Coleman and Hodges, 1995). Rainfall distribution and weathering patterns were modified that in turn influenced carbon dioxide fluxes, in general reducing them. In addition, the rising of the Panama Isthmus blocked North Atlantic surface waters from reaching the Pacific Ocean, deflecting them northward to affect the North Atlantic drift and oceanic bottom waters, and in turn, the climate of northern European countries (Crowell and Frakes, 1970; Coates et al., 1992; Jackson et al., 1993; Vermeij, 1993; Stanley, 1995; Collins et al., 1996; Ross, 1996; Burton et al., 1997; Haug and Tiedemann, 1998). In fact, this example of the closing of a gateway between continents, and the rerouting of ocean-current flow, is a main conclusion of this review: Continental arrangements and their influence on heat distribution in the air-ocean system are quite significant in locating climate zones and in causing climate change.

From the late Eocene (ca. 43 Ma) one of the warmest intervals in Earth history prevailed back to ca. 115 Ma throughout the Late Cretaceous and well into the Early Cretaceous (Frakes et al., 1992; Francis, 1993). This warm interval was preceded by two (perhaps three) *Mesozoic Cool Intervals* when ice-rafted debris reached ocean shelves. On the whole, however, the Mesozoic witnessed warm climates compared with both the late Paleozoic and the late Cenozoic. During the approximately 213 m.y. time span between the beginning of the Late Cenozoic Ice Age and when the Late Paleozoic Ice Age ended at ca. 256 Ma, the climate was cool enough for ice near shore but there is no direct record of glaciers on land. These icy intervals occurred in the Early Cretaceous (between ca. 115 Ma and ca. 140 Ma, but perhaps as late as 105 Ma), questionably in the Late Jurassic (between ca. 160 Ma and ca. 175 Ma), and in the Early Jurassic (between ca. 188 Ma and ca. 195 Ma). For the ca. 60 m.y. or so from these times back to just before the end of the Permian period, warm climates predominated, including all of Triassic time.

In the warm Late Cretaceous no direct record of even river or shore ice has been reported, so that an explanation for sea-level fluctuations is enigmatic (Stoll and Schrag, 1996). The Late Cretaceous is viewed as a time of relative meridional equitability in climate because the existence of dinosaurs and vegetation at northern high latitudes suggests mild seasonal temperatures (Brouwers et al., 1987; Parish et al., 1993). From southern high latitudes, low temperatures and biologic diversity are reported (Rich et al., 1988; Gregory et al., 1989). Computer model studies of Late Cretaceous climates suggest that latitudinal contrasts and seasonality contrasts permitted cold winters, with some ice here and there, and quite warm summers (Barron and Washington,

1984; Crowley et al., 1986; Barron and Peterson, 1989). Rhythmic bedding with a short time frequency has been credited to orbital changes (R.O.C.C. Group, 1986; Herbert and Fischer, 1986). If there were glaciers during the Late Cretaceous, where were they?

During the Late Cretaceous and early Tertiary, plate reconstructions suggest that any deep basinal sites at high latitude and at locations where glacial debris might record nearby glaciers on land still remain at depth. Such sites have not been tectonically uplifted and so are not available for investigation (Scotese, 1990). Incipient opening of the North Atlantic Ocean may have begun at the end of the Cretaceous at about 50° N lat and along the Labrador Sea with high rift shoulders in Greenland and Labrador (Scotese, 1990, fig. 15). Any stratal record of glaciers, if present, remains deeply buried and is overlapped by younger beds infilling North Atlantic regions and the Labrador trough as they widened. At high southern latitudes rift sites are shown where Australia broke away from Antarctica at ca. 84 Ma (Scotese, 1990, fig. 17). I speculate that glaciers in these regions may have waxed and waned to bring about Cretaceous sea-level fluctuations, but the stratal record is deeply buried and so far unavailable to us.

Lower Cretaceous strata of both Australia and Siberia contain evidence of ice rafting during an *Early Cretaceous Icy Interval* but it is not yet clear whether there was true glacial ice upon nearby continents, although local ice caps are likely (Frakes and Francis, 1988; Frakes and Krassay, 1992; Frakes, et al., 1995). In northern South Australia, within the southeastern part of the Eromanga basin, dropstones of Precambrian quartzite and volcanic rock up to 3 m in diameter, interpreted as ice rafted, lie within dark bioturbated or laminated sandy shale of the Bulldog Shale (Fig. 2). Glendonites, indicating water temperatures below 5 °C, and quartz sand grains with characteristic glacial surface textures, occur along with occasional wood fragments. In the central Carpentaria basin of the Northern Territory and adjoining Queensland, ~1,000 km to the northeast of the Eromanga exposures, similar deposits carry dropstone boulders with rare penetration structures into underlying substrate (Frakes et al., 1995). During the Early Cretaceous, the Eromanga deposits lay between 60° and 70° in southern latitudes and the Carpentaria deposits at around 55° (Embleton, 1984). Biostratigraphic control does not yet fix the age or synchroneity of these deposits satisfactorily, but they occur within the approximate range ca. 115–140 Ma. It is also not yet documented whether the ice was river or shore ice, or whether glaciers on land were the source.

During the Early Cretaceous, Australia and Antarctica were joined, and both lay at high southern latitudes (Rich et al., 1988; Stump and Fitzgerald, 1992). Rift shoulders high enough for glaciers existed in the Ellsworth Mountains of West Antarctica, with a relief of at least 2 km based upon fission-track studies (Fitzgerald and Stump, 1991). The rifting was associated with the beginning of the separation of East and West Gondwana, and the opening of the Weddell Sea. New Zealand was also attached to Antarctica and Australia at that time, whence diamictites of Aptian age (ca. 110 Ma) have been reported (Waterhouse and

Flood, 1981). These New Zealand layers, including conglomerates, require study to work out their depositional environment and emplacement processes.

In the Northern Hemisphere, ice-rafted material is reported from Lower Cretaceous deposits in Siberia between the Verkhoyansk region and Kamchatka (Epshteyn, 1978). Megaclasts are also recorded within Lower Cretaceous strata of the Canadian Arctic, Spitzbergen, the northern Brooks Range of Alaska, and possibly the Songliao basin of China (Kemper, 1983; Frakes and Francis, 1988). These localities all lay above 60° N lat. Again, river and shore ice document a cold climate, but there is no direct evidence of inland glaciers. Interpretations of the stratigraphic record in Germany and elsewhere in the Northern Hemisphere suggest that two cold episodes occurred during the Early Cretaceous, one in Aptian-Albian time (ca. 105–115 Ma) and the other

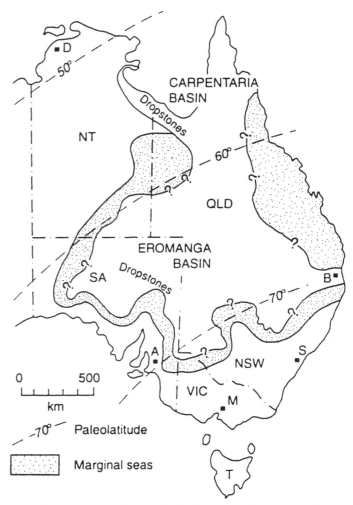

Figure 2. Eastern Australia during Early Cretaceous (Aptian) time showing Carpentaria and Eromanga basins. Dropstones occur in strata of marginal seas where labeled. QLD = Queensland; NSW = New South Wales; NT = Northern Territory; SA = South Australia; T = Tasmania; VIC = Victoria; A = Adelaide; B = Brisbane; D = Darwin; M = Melbourne; S = Sydney. After Frakes et al. (1995, fig. 1); includes paleolatitudes from Embleton (1984).

in Valanginian time (ca. 130–140 Ma). Strontium and oxygen isotopic studies suggest sea-level fluctuations of as much as 50 m during the Berriasian and Valanginian. These are interpreted as due to inland glaciers waxing and waning, perhaps in Antarctica (Stoll and Schrag, 1996). The fluctuations match the Exxon sea level curve satisfactorily. In fact, "the validation of rapid changes in the sequence stratigraphic sea level curve for the Early Cretaceous implicates glaciation as a possible explanation for all such sea level changes" (Stoll and Schrag, 1996).

The next older warm interval during the Late Jurassic (between ca. 140 Ma and ca. 180 Ma) may have been interrupted briefly by a possible short time of iciness between ca. 160 Ma and ca. 175 Ma (Frakes et al., 1992; Eyles, 1993). Earlier, ice-carried debris indicates a *Middle Jurassic Cold Interval* during the Bathonian stage (ca. 160–168 Ma) (Moore et al., 1982; Chumakov and Frakes, 1997). In northern Siberia pebbles and rare cobbles in muddy argillite beds are interpreted as laid down by shore or river ice in a shallow marine location, but there is no direct evidence of glacial ice in the Jurassic hinterland at this time. Sea-level variations of older Jurassic beds in southern Germany, assigned to the Sinemurian and Pliensbachian stages (ca. 188–200 Ma) have been attributed to the waxing and waning of distant glaciers (Brandt, 1986). Future work upon both sea-level fluctuations and paleoclimate indicators may therefore document Jurassic cool episodes when glaciers expanded.

Strong earlier ice ages when glaciers flourished upon continents are documented for the late Paleozoic, during the Devonian-Carboniferous and Ordovician-Silurian transitions, for several times in the Neoproterozoic, for the Paleoproterozoic, and in the mid-Archean. Description of these pre-Mesozoic ice ages is the purpose of this treatise. The glacial record, so far as now known, goes back about 3 Ga and began at a time not long after the assembly of continental cratons.

RECOGNITION AND PRESERVATION OF ANCIENT GLACIAL RECORDS

The ultimate test of science is in the observations.
Oliver, 1996, p. 1

*The rock record, . . . not computer models, is
the primary source of data on ancient climates.*
Eyles, 1993, p. 211

The veracity of an ancient ice age depends on both direct and indirect evidence fitted into a satisfactory paleogeographic reconstruction. Because we live today when ice sheets and large glaciers thrive on Earth, as on Antarctica and Greenland, geologists can investigate glacial products and ascertain what kind of record will remain after the ice has melted (Frontispiece). Ice sheets and tongues now flow from inland areas and reach outward to the open sea and have recently retreated since Pleistocene times (Denton and Hughes, eds., 1981). They have left behind distinctive landforms and scourings on the bedrock and have laid down characteristic sedimentary facies, both on land and within the sea

(Goldthwait, 1971, 1975; Jopling and McDonald, 1975; Davidson-Arnott et al., 1982; Anderson, 1983; Andrews and Matsch, 1983; Ruddiman and Wright, 1987; Wright, 1989; Dowdeswell and Scourse, 1990; Anderson and Ashley, 1991; Brodzikowski and van Loon, 1991; Huggett, 1991; Parish et al., 1993; Miller, 1996; Menzies, 1995, 1996). Floating icebergs carry rocky fragments far from shore where they melt and drop identifiable glacial debris upon distant ocean floor. Such records, preserved in ancient strata both on land and in marine deposits, provide direct evidence of previous glaciation. An indirect record comes from the interpretation of sea-level changes and from sea-floor cores and ancient stratal sequences (e.g., CLIMAP Project Members, 1976, 1981; Wilgus et al., 1988; Molnia, 1983a). Geochemical variations such as changes through time in ratios of isotopes, especially those of oxygen, carbon, sulfur, and strontium, may follow upon occurrences of continental ice (Holland, 1984; Faure, 1986; Gregor et al., 1988; Miller et al., 1996b). The geochemical proxies of temperature, and through them of climate, substantiate past cool and icy intervals.

Direct evidence of past glaciation

Investigations of modern and late Cenozoic deposits and bedrock floors beneath and at the margins of glaciers, or where glaciers have but recently disappeared, provide clues for interpreting times and places of past glaciations (Embleton and King, 1975a,b; Drewry, 1986; Sugden and John, 1976; Brodzikowski and van Loon, 1991). Glaciers scour into bedrock cutting U-shaped valleys and manufacturing roches moutonnées and stoss-and-lee structures. Indurated bedrock surfaces are striated and veneered with glacial polish (Figs. 3, 4). These forms and features are distinctive and identifiable at many places in both modern and ancient environments and guide investigators of ancient ice ages as they put together paleogeographic reconstructions. For many ancient ice ages, however, details of sedimentary facies transitions from glacial margins to far offshore are scarce; they have been eroded or are covered by younger deposits. At best, however, two-dimensional organizations of glacial and periglacial facies are complex, and three-dimensional depictions even more so (Figs. 5, 6).

Glacial striations are distinctive if well preserved (Embleton and King, 1975b) and occur on both floors beneath glaciers and on stones and blocks picked up and carried by them. The striations occur in sets with slightly divergent directions, associated with chattermarks, crescentic gouges, lunate fractures, and other types of friction features. Glacial landforms with friction features have been identified someplace on Earth for all of the major ice ages of the past except Archean. For all ice ages, however, cobbles and boulders with glacial facets and striations have been recovered from stratal deposits and provide strong evidence of past ice action. Scratches of other origins, such as those from fault slickensides or grinding within debris flows or from bolide impacts, can usually be distinguished from glacial striations.

Glacial sedimentary products. In lakes, seas, or oceans marginal to glaciers, and even in distant offshore regions, two

Figure 3. Polished and striated glacial pavement, Nooitgedacht Preserve, Kimberley District, South Africa. Permo-Carboniferous glaciers have left their record upon Precambrian Ventersdorp Lava. Scale is 15 cm long. February 1967. Republished from Crowell and Frakes (1972, fig. 10).

Figure 4. Permian glacial striations and grooves upon indurated Devonian Furnas Sandstone, Colônia Wittmarsum, Paraná State, Brazil. July 1988. A. C. Rocha-Campos in photograph.

distinctly different stratal facies result after the glacial ice has melted: (1) massive unsorted debris, and (2) the dropstone facies where debris and especially large stones and boulders have been dropped from ice floating in the sea, an origin most easily identified in thin-bedded sequences. As glaciers grind along, sedimentary debris accumulates on them, within them, and at their bases, and commonly consists of unsorted clasts dispersed and mixed into clay, silt, and sand. Even if the ice has disappeared, if a glacier is still present or if Quaternary glacial landforms are preserved in the vicinity, the glacial origin of the unsorted debris is probably confirmed.

In ancient stratal sequences, however, because ice has long since melted and landforms have eroded away, it may be difficult to designate a glacial origin because similar unsorted material may form in many ways (Crowell, 1957; Dott, 1961, 1963; Friedman and Sanders, 1978; Hambrey and Harland, 1981; Koster and Steel, 1984; Eyles, 1993; Miller, 1996). Evidence other than the mere presence of an interlayered bed of unsorted debris is required before a glacial origin for the deposit is acceptable. Descriptive terms for such unsorted rock debris, with no connotation regarding manner of origin, are *diamicton* where the material is unconsolidated and *diamictite* where indurated (Flint et al., 1960; Schermerhorn, 1966; Hambrey and Harland, 1981; Dreimanis, 1983; Visser, 1983a,b).

Tills and tillites are diamictons or diamictites of identified glacial origin consisting of unsorted and poorly mixed rock debris with grain sizes ranging upward from fine clay to dispersed huge boulders and blocks (Figs. 7–10). Such glacigenic material may be laid down beneath a glacier to form lodgment till, or deformation till if it displays folds and faults and isolated deformed bodies, or debris concentrated where it sinks through ice as it melts (melt-out till) or where it moves laterally down the terminus of a glacier or down a slope in front of the glacier (Boulton, 1972; Boulton and Deynoux, 1981). These till types may be difficult to identify in the ancient geologic record in the absence of geomorphic information on the location and shape of the contributing glacier that has now disappeared. The products of transitional processes where till moves farther downslope as debris flows or within turbidity currents are particularly difficult to identify (Frakes and Crowell, 1967; Schermerhorn, 1974, 1976; Young, 1976a).

Massive diamictites commonly contain faint wispy interlayers, clots of coarser material, and even littoral fossils indicating downslope sliding from shallow to deeper water (Molnia, 1983a; Anderson and Ashley, 1991). Only where such facies, formed by downslope sliding and mixing of unconsolidated mud with interbedded conglomerate and twisted rip-up fragments, are known to grade laterally into deposits of glacial origin, is a glacial origin for some of the debris acceptable. Otherwise, the debris may have been emplaced by downslope sliding alone without input from glaciers, and several different movement processes may be identifiable in the deposit (e.g., Lowe, 1979). Glacially faceted and striated megaclasts within the debris may establish, however, that glacial sources existed someplace upflow from the

Glacial Environments

Supraglacial subenvironments

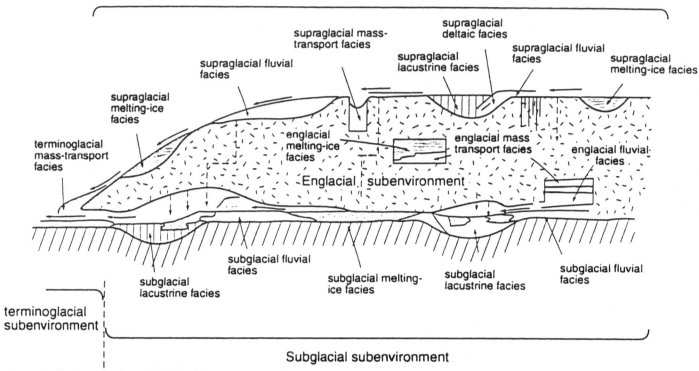

Figure 5. Sketch of continental glacial environments, showing the three main subenvironments of ephemeral sedimentation: supraglacial, englacial, and subglacial. After Brodzikowski and van Loon (1991, fig. 63).

Marine periglacial environment

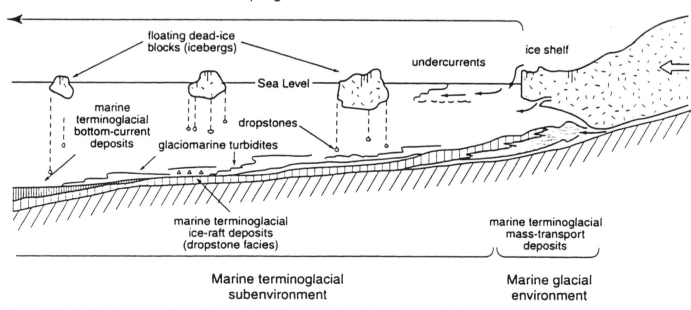

Figure 6. Sketch of environments where a floating ice shelf from a continental glacier spawns icebergs where it reaches the sea. After Brodzikowski and van Loon (1991, fig. 64).

Figure 7. Permo-Carboniferous Dwyka tillite at Port St. Johns, Cape Province (Transkei), South Africa. Looking eastward toward source region that is now occupied by the deep Indian Ocean. Tillite includes subangular to subrounded boulders up to 90 cm in diameter of coarse yellow granite, dark argillite, quartzite, quartzitic sandstone, and gneiss. The dispersed till-fabric direction is toward the southwest. February 1967. Republished from Crowell and Frakes (1972, fig. 5).

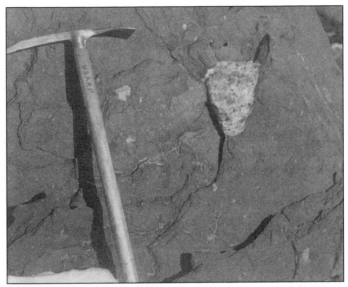

Figure 9. Massive tillite, Whiteout Conglomerate, Permo-Carboniferous. Near Mt. Earp, Sentinel Range, Ellsworth Mountains, Antarctica. USARP = U.S. Antarctic Research Program, primarily supported by the U.S. National Science Foundation. December 1966.

Figure 8. Massive Neoproterozoic Marinoan tillite, Elatina Formation, Umberatana Group, Adelaidian Sequence, South Australia. Brachina Creek section, near Oraparinna Mine, Flinders Ranges. August 1976.

Figure 10. Bouldery Dwyka tillite, Permo-Carboniferous. Recent studies show that the boulder bed contains mudstone partings defining bedding thicknesses of 1–1.5 m. Inclined dips of large clasts suggest a debris-flow origin for lower beds that were then overridden by ice (J. N. J. Visser, 1998, written commun.). Elandsvlei, Cape Province, South Africa. February 1967.

resting place of the deposit, but at an unknown distance, and that the glacial contribution may have been quite small.

Large stones (megaclasts) may be useful in distinguishing between downslope sliding and glacial processes, or in appraising the proportion of the two contributing processes. They may be carried by glacial ice for great distances without being further worn or abraded, and so may be angular and mixed within an unsorted matrix (Boulton, 1978; Miller, 1996). Boulders may consist of rock types not known to exist in nearby regions or in areas within reach of fluvial or other methods of water transport. In ancient deposits, if

distant sources for megaclasts are recognized or suspected, ice as the transporting agent needs evaluation. Glacial megaclasts may have arrived at their final resting place by nonglacial agents, or moved downslope by sliding and roiling or within turbidity currents, and constitute only a very minor proportion of the debris deposited.

Some diagnostic features are found on striated pavements or within glacial till laid down beneath glaciers or very close to their margins (Figs. 11, 12). At places within modern glaciers, boulder or clast pavements are manufactured; these arc also identified in ancient deposits (Figs. 13, 14). They consist of trains of large clasts frozen within till that are held fast so that their tops are ground off by ice shod with rocky debris moving above a shear surface (Holmes 1941; Visser and Hall, 1985; Clark, 1991). The overlying ice contains frozen till that grinds along the shear surface at the base of the overgliding mass. As a result, megaclasts are faceted, polished, and striated with the striations usually aligned from stone to stone and with many stones displaying a beveling on their upflow margins. Boulder pavements have been identified for most of the ancient ice ages. Some cobbles and boulders display "flat-ron" shapes; others are shaped with blunt rounded ends remindful of bullets, hence "bullet-boulders" or "bullet-stones." Some of these have striations curving around their upstream rounded and polished ends and with flaring striations along their sides (Fig. 15).

Other features are helpful in documenting a glacial origin for diamictites laid down at or near a glacier, such as lodgment till or tillite. Here the finest material is commonly indurated rock flour, composed of rocky material mechanically ground down to silt and clay. Under a petrographic microscope, broken sand and silt grains document mechanical breaking such as would occur within a glacial environment. Only a few studies have been undertaken to show that the finest material is the product of mechanical weathering in contrast to deep chemical weathering, however, and to show that the material consists of consolidated glacial rock flour (Deynoux, 1985). In outcrop, many tillites with a fine clay matrix and widely dispersed stones display conchoidal fracturing and spheroidal weathering. Other distinctive facies include irregularly bedded diamictites (Figs. 16, 17) and periglacial structures such as sandstone wedges formed when ice has melted from deep fissures characteristic of patterned ground in permafrost regions and then infilled by drifting sand. These have been reported and are illustrated from upper Proterozoic strata in Mauritania (Deynoux, 1982) and in Australia (Williams and Tonkin, 1985). During Precambrian times a different balance may have prevailed between mechanical and chemical weathering processes on continents so that research is needed to interpret the process from the preserved resulting product (Young and Nesbitt, 1985, 1999).

Field measurement of stone fabrics may be helpful in disclosing the patterns and directions of glacial flow in a lodgment tillite (Holmes, 1941; Drake, 1971, 1972; Deynoux, 1985). Ideally, such studies should involve the measurement of long axes of elongate and flat stones to show flow imbrication by means of a three-dimensional plot. Because extracting and measuring the orientation of such clasts from indurated strata are difficult, however, more often a two-dimensional concept only of the flow is obtained by measurements of long axes of stones on bedding planes (Frakes and Crowell, 1967; Stratten, 1969).

Mineral grains carried by moving ice and pressed firmly

Figure 11. Permian glacial striations with crescentic gouges at Hallet Cove, Fleurieu Peninsula, South Australia. Ice moved from upper right to lower left over Precambrian basement. Conspicuous striations in center of photograph trend N 40° W ± 10°. The crescentic gouges are somewhat younger and trend N 10° W ± 10°. Centimeter scale. May 1968.

Figure 12. Glacial striations with large crescentic gouge beneath Permo-Carboniferous Dwyka tillite, scribed upon Precambrian Ventersdorp Lava. Nooitgedacht Preserve, Kimberley District, South Africa. Pointed scale, ~15 cm long, shows direction of ice flow toward south-southwest. February 1967. Republished from Crowell and Frakes (1972, fig. 13).

Figure 13. Boulder pavement within Permo-Carboniferous Dwyka tillite. Elandsvlei, Cape Province, South Africa. Ice moved in direction of pointed scale (15 cm long), toward east-southeast, beveling upflow ends of some boulders, and faceting and striating them so that they show uniform orientation of striations. February 1967. Republished from Crowell and Frakes (1972, fig. 11).

Figure 14. Side view of a boulder pavement extending into the outcrop upon the upper surfaces of the aligned stones. Permian Itararé Subgroup, railway cut near Jurumirim, São Paulo, Paraná basin, Brazil. Knife is 8.2 cm long. June 1975.

against each other or against bedrock are at places fractured and scarred with percussion markings and etchings. Quartz grains under the electron microscope may display marks characteristic of glacial transport (Krinsley and Doornkamp, 1973; Mahaney, 1995). Garnet grains viewed with a microscope may show trains of chattermarks, probably also caused by glacial transport (Folk, 1975; Gravenor, 1979, 1985). Criteria are still needed, however, to distinguish between glacial percussion marks and chemical etching marks formed during diagenesis (Gravenor and Leavitt, 1981). Unfortunately, such markings are worn away as quartz and garnet grains are carried by fluviatile and other transporting agents after they have left the proximity of the glacier; thus their usefulness as environmental indicators is limited.

Figure 15. Two bullet-shaped cobbles, of which the larger one is ~7 cm long. The bullet-cobbles have rounded blunt ends with faint glacial striations, interpreted as formed within an anchored ice-flow regime, but have later been rolled about in coming to rest within the sandy tillite. Permian Talchir tillite, Gungata River section, ~10 km southwest of Ambikapur, Son-Mahanadi basin, Madhya Pradesh, India. January 1981.

Figure 16. Transitional periglacial facies, irregularly bedded. Dwyka Tillite transition. 35 km northwest of Loeriesfontein, South Africa. Photograph August 1982 of slab collected by J. N. J. Visser.

Dropstone facies. Where glaciers reach the sea or a large lake, at least two distinctive facies occur: the massive-bedded dropstone facies and the thin-bedded dropstone facies (Figs. 18–20). The massive-bedded facies occurs where ice-rafted glacial debris falls into an environment where large quantities of silt and clay accumulate without significant laminations or bedding. Lateral currents are minimal. Such environments commonly lie just seaward of wasting ice fronts contributing huge volumes of rock flour to milky glacial streams debouching into a protected nearshore site (Ojakangas, 1985). The clay settles to the sea floor to form featureless deposits along with isolated stones and clots of unsorted coarse debris dropped into accumulating mud.

Thin-bedded dropstone facies occur where ice-rafted stones and debris fall into a quiet depositional environment as shown by the association with thin beds or laminations in fine-grained strata. Rafting may be confidently inferred because of the contrast in grain size and inferences concerning the hydrodynamic processes involved. Laminations result from very weak or no lateral currents with deposition occurring in quiet water. Coarse debris can be carried laterally only by strong currents. Rafting is inferred where large clasts depress or puncture the bedding or laminations, bending them downward, and where there is no evidence of sufficiently powerful lateral currents to carry the larger stones. The substrate may even have been deformed as stones fell and penetrated it, but it may be difficult in outcrop to distinguish dropstone impact from bending around stones during later compaction. Isolated stones, known as "lonestones" descriptively, may be genetically designated as "dropstones" when rafting is shown as the manner of transportation and emplacement.

In such thin-bedded and fine-grained sequences, a dropstone origin is best inferred where the diameter of the stone is many times, perhaps 10 or 15 times, the thickness of individual laminations. Confidence in this dropstone interpretation is enhanced where the stones are truly alone, and penetration structures identified. Where a flat stone has its plane of flatness nearly perpendicular to the bedding, it may be inferred that the stone punctured the bedding as it fluttered downward through the water column to drop into place. Where icebergs tilt or melt and dump unsorted debris upon the depositional floor, the material is usually unsorted and occurs in clumps or patches as the result of iceberg melting or overturning (Fig. 21) (Thomas and Connell, 1985). The deposit may also contain pea-sized "till pellets" made up of clay where rock flour carried in fresh water became flocculated upon meeting electrolytes in ocean water (Ovenshine, 1970).

If gravel, sand, and stones are strewn out along a bedding surface within a fine-grained sequence and are mixed with unsorted sedimentary debris, careful scrutiny is required to eliminate lateral transporting currents from consideration. Isolated lonestones may not be dropstones. For example, currents washing to a site laterally, especially by vigorous turbidity currents, may bring in large clasts and then move on, carrying away the finer material and leaving behind only the larger stones. These lonestones are known as "lagstones"—they lagged behind and were left behind by the current. Current bedding or small-scale cross stratification and

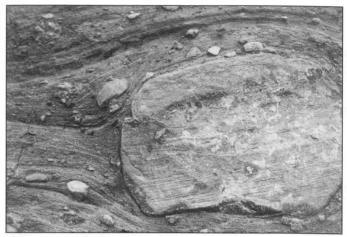

Figure 17. Lonestone in tillitic transitional facies into subaqueous periglacial environment. Neoproterozoic Sinian series, Luoquan Formation, Shangzhanquan locality, Lonan County, Xi'an District, Shanxi Province, China. Large stone is ~10 cm in diameter. October 1981.

Figure 18. Dropstone boulder, dropped into thin-bedded strata. Permian Tubarão Group, Itararé Subgroup. Pedreira Quarry, Paraná, Brazil. July 1988. John F. Lindsay in photograph.

Figure 19. Dropstone within thin-bedded strata. Permian Dwyka tillite, transitional facies into subaqueous environment. Quarry, 10 km northeast of Vryheid, Natal, South Africa. Scale in centimeters on right, inches on left. July 1970.

Figure 21. Cluster of unsorted glacigene debris upon bedding plane, inferred to have been dumped when a debris-laden ice floe or iceberg overturned or melted. Permian Shoalhaven Group, Wasp Head Formation. Wasp Head, southeastern Sydney basin, New South Wales, Australia. Knife is 9 cm long. August 1986.

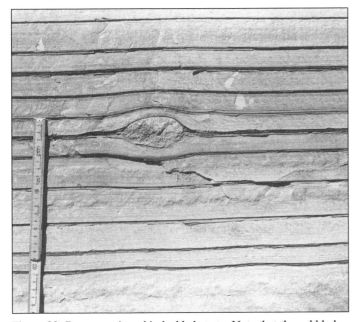

Figure 20. Dropstone into thin-bedded strata. Note that the pebble has fallen upon shale, laid down after deposition of the fine-grained sandstone layer beneath it, and could not have been carried to its resting place by the weak currents responsible for the lower sandy layer. In addition, it is overlain by mudstone showing that it is not a basal clast of the overlying sandy layer. Permian Itararé Subgroup, Itu Formation. Itu Quarry, near Itu, São Paulo, Brazil. Scale in centimeters. November 1972.

other sedimentary structures may provide clues indicating such lateral currents, and make suspect the interpretation that the lonestones are truly dropstones. Volcanic eruptions and bolide impacts may also throw coarse debris or megaclasts high aloft, which then fall to resting places as dropstones. These lonestones may be termed "blownstones" or "flownstones" or "thrownstones," and need discrimination from those dropped by melting ice or other rafting mechanisms mentioned below.

Dropstones may also fall from floating rafts other than ice, such as stones entrapped in the root mats of floating trees ("rootstones"), those that serve as holdfasts for kelp ("kelpstones"), or those used by sea lions or other sea mammals for crushing shells and then tossed aside ("sealstones") (Emery, 1955, 1963). In summary, thin-bedded and laminated sequences demand careful scrutiny for dropstones if glaciation within the region is suspected. And if dropstones are indeed identified, sources other than wasting ice need appraisal. In interpreting the dropstone record of the remote past, judgment is therefore required about whether the ice came from nearby or distant sources, a judgment depending on the reliability of the paleogeographic and ocean current reconstructions. Reconstructions for limited time slices of the ancient past are only now becoming useful (McKerrow and Scotese, 1990; Scotese, 1990).

Ice floating in the sea, however, does not necessarily come from true glaciers. River ice and shore ice may freeze around rocky debris and carry this material to a distant quiet-water site. Floes of river and shore ice are commonly not heavy or thick enough to either manufacture facets and striations or float large megaclasts to reach the open sea. Large icebergs calved from glaciers, however, may float massive blocks several meters in diameter (Gilbert, 1990).

Along the coast of the Gulf of Alaska today, tongues of alpine glaciers debouch directly into deep marine fjords that open to the Pacific Ocean (Molnia, 1983b; Eyles and Lagoe, 1990; Eyles et al., 1991). Icebergs float away and carry glacial debris to distant sites. In the modern Arctic and North Atlantic regions, bergs from sources along the Siberian coast have been identified as far south as the Grand Banks off Newfoundland (Fig. 22) (Crowell, 1964; Lewis and Keen, 1990). Siberian clasts are carried by currents westward along the margins of the Arctic Ocean,

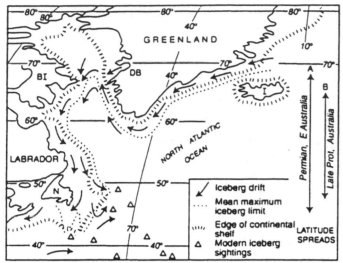

Figure 22. Iceberg drift paths in North Atlantic Ocean today. Based on U.S. Defense Map Service Chart, 1990. BI = Baffin Island; DB = Disko Bay, Greenland; I = Iceland; N = Newfoundland. On right, latitude spreads for permitted ice-rafted debris in Australia: A = Permian of eastern Australia, and B = Late Proterozoic of central Australia. (See text.)

southward along the east coast of Greenland, and then north again after rounding Cape Farewell at the southern tip of Greenland. In Davis Strait and the Labrador Sea they join bergs from Disko Bay in west Greenland and then float southward more than halfway to the equator to melt and dump their debris on and near the Grand Banks off Newfoundland. Bergs have been sighted recently as far south as 32° lat (USDMA Chart, 1990) and no doubt drifted even farther south during Pleistocene glacial culminations, especially during episodes of massive discharge during ice sheet surges and Heinrich events (Hollin, 1969; Heinrich, 1988; Bond et al., 1992; Grimm et al., 1993; Oerlemans, 1993; Broecker, 1994; Hunt and Malin, 1998). If such debris were found in ancient strata, the inference that the nearest land—in this modern Grand Banks case, Newfoundland and Labrador—was the site of parent glaciers would be quite erroneous. This example emphasizes that *ice-rafted debris within a stratal section reveals little concerning the location of glacial centers.* We need to know the arrangement of ocean currents as well. *Such identified glacial material tells us only that glaciers existed somewhere upcurrent.* For times in the ancient geologic past, however, reconstructions of land and sea arrangements and the organization of ocean currents are seldom reliable.

Glacier sites and facies transitions. Sites for glacial growth primarily include uplands or mountains and especially those along continental margins. Glaciers also occur upon uplands well within continents and within broad interior depressions so that distinctive glaciogene products will not reach the ocean. During the Late Cenozoic Ice Age, for example, the Laurentian Ice Dome developed over a broad depression within the North American craton, now occupied by Hudson Bay (Denton and Hughes, eds., 1981). The Scandinavian Ice Sheet grew over part of the Gulf of Bothnia. The Greenland Ice Cap lies within a large

saucer-shaped depression surrounded by restraining mountainous islands. Quite a different site for ice accumulation is recognized where north-flowing rivers entered the Arctic Ocean along the Siberian coast during the Pleistocene. At times river mouths in this region become choked with stacked river-ice floes, which may grow to form glacial centers as more and more cold water flows polarward to these barriers and then freezes (Denton and Hughes, 1981b). Many of these types of late Cenozoic sites have not yet been recognized within the ancient record. Neither shallow depressions nor uplands are likely to survive erosion through long periods of geologic time.

Within a few kilometers downflow from a continental glacier, no vestige of the glacial origin may be detectable because the glacial material has been reworked, carried, and deposited by other transporting agents (Figs. 23, 24). Textures and sedimentary structures reveal instead the new mode of transportation and deposition after the material has been carried away from the glacial front. Till lying upon a glacially scoured pavement will have given way to water-carried deposits lying unconformably upon basement or within a nonglacial sedimentary sequence. All local evidence of glaciers, marginal or at distance to the depositional site, may have disappeared. Even the inland home of the glaciers may have been worn away. Erosional features manufactured early during the advancing stages of an ice age are also destroyed as glaciation continues. A preserved record of the earlier parts of the glaciation is unlikely (Crowell, 1978). The record available today probably formed during glacial retreat with striated surfaces covered by regressive deposits.

Dropstones that appear within a stratigraphic section laid down at great distance from an inferred glacier may have been deposited a very long time after the ice cap began to grow on land. Time elapsed as the glacier grew to reach the sea. There will be no record of glaciation in the stratigraphic section in the subsiding basin during this time of distant glacial growth. An evolutionary model where ice caps grow on continents implies that the preserved record in a distant basin will lack documentation for both the early stages of glacier growth and the ending stages. In reconstructing a paleogeographic model across a sweep from continental interiors to deep basins, as emphasized in Figures 23 and 24, we must ponder these concepts. Although it is distasteful to advocate a glacial input where there is no direct evidence for glaciation, the logic of such a growth and waning model through time requires assessment. Fortunately, in some cases geochemical and paleontological data may document a cooling in the inferred preglacial beds followed by a warming in the inferred postglacial beds. Inferences from the indirect record of climate cooling and iciness followed by warming may come to our rescue.

Glaciers are therefore adequately documented only where a facies reconstruction fits together many different types of evidence from several correlated sections and where there is confidence in lateral and time correlations. Facies studies undertaken during recent years are now improving paleogeographic interpretations for several of the ancient ice ages (e.g., Moncrieff and Hambrey, 1990; Anderson and Ashley, 1991; Eyles, 1993; Link

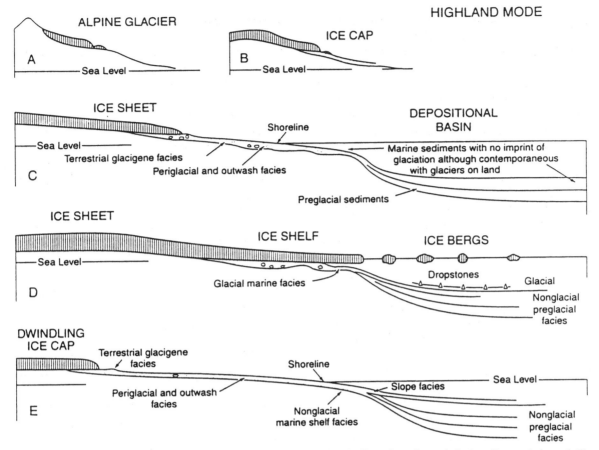

Figure 23. Facies relations associated with growth, culmination, and dwindling of continental glaciers (Stages A through E): Highland mode. Alpine or upland glaciers grow to form an ice cap that enlarges through time to make an ice sheet. Strata preserved in the basin containing a discernible glacial record were deposited long after the glaciers began to grow on land and are termed *preglacial*, that refers to time and stratigraphic position. They may have been deposited at the same time as growing glaciers at distance. The term *preglacial* needs differentiation from *periglacial*, which refers to environments peripheral or adjoining a glacier. Proglacial refers to position in front of glaciers. Postglacial refers to time after the glaciation. (See text.) From Crowell (1983a, Fig. 1).

et al., 1994; Proust and Deynoux, 1994; Von Brunn, 1994; Visser, 1997a, b). Correlations are also improving because of distinctive geochemical and paleomagnetic and sea-level excursions from the norm. For ancient times the record of punctuated events such as bolide impacts, volcanic eruptions, rapid eustatic changes in sea level, and paleomagnetic and geochemical digressions are quite short geologically and reach over large areas (Christie-Blick and Driscoll, 1995; Christie-Blick et al., 1995). They provide hope for improved correlations over huge regions as the products and consequences of these events are recognized, sorted out, documented, and related to interpretations from sequence stratigraphy and isotopic excursions.

From the pre-Mesozoic viewpoint of this review, paleogeographic reconstructions for the Late Cenozoic Ice Age and Mesozoic times of iciness are mainly adequate and provide a basis for paleoclimate reconstructions. Within the Recent, correlations in time are usually satisfactory to within a few hundred or a few thousand years, and in the early Pleistocene, to several thousand

years. Such correlations are extended with confidence over very large regions and even worldwide. For the late Paleozoic and Late Proterozoic ice ages, correlations closer than a few hundred thousand years, or more than a million years, and over regions of subcontinental size are rare in the absence of very special time markers, such as correlatable dated ashbeds and perhaps distinctive isotopic excursions.

In making paleogeographic reconstructions, the dated time ranges of depositional and tectonic events must not exceed those assigned to the time limits of the reconstruction. If correlation across a region depicted cannot be closer than, say, 5 m.y., the depiction must not show details that would change significantly within this time span. During 5 m.y. shorelines may move great distances laterally and mountain ranges may rise and be worn away. Such reconstructions should therefore be shown only as diagrammatic sketches. During recent decades geologists have come to recognize that the Earth's crust has long been mobile as the concept of plate tectonics has been extended back to when continental cratons were first assembled (Bickford, 1988; Hoffman, 1989a,b, 1991; Windley, 1993; Rogers, 1996). The likelihood of crustal mobility throughout geologic time raises the likelihood as well that the topography and the waxing and waning of glaciers have changed commensurately. In short, the locations of the glaciers themselves depend on the acceptability of paleogeographic reconstructions that in turn depend on regional information known to be synchronous within the bounds of correlations and dating. Such reconstructions are reasonably satisfactory for Phanerozoic time and much less satisfactory for late Neoproterozoic time, but they are woefully inadequate for earlier times.

In southern Africa, details of facies deposited during the Late Paleozoic Ice Age are quite complete and well exposed in the semiarid climate, and are far more accessible than those during much of the late Cenozoic record, especially the marine record. Facies changes can be "walked out" by field geologists whereas much of the marine late Cenozoic record lies beneath the sea and interpretation depends on correlation from core to core and by means of geophysical profiling. In addition, in southern Africa and southern South America, a record of both the beginning and the end of the ice age is preserved so that we can infer what brought about these events. For the Late Cenozoic Ice Age, however, we are still living within it and it has not yet ended.

The recognition of intervals of climate coolness, including times of strong and widespread glaciation, fortunately are supplemented by indirect evidences, briefly described next. The direct record, however, provides essential calibration of glacial episodes with that of the indirect record.

Indirect record of past glaciations

Ancient stratal sequences reveal the rise and fall of sea level that may be the result of the waxing and waning of ice sheets; isotopic and other geochemical variations preserve a record that may be interpreted to show climate warmings and coolings. The geochemical proxies of oceanic temperatures are especially useful because they are largely independent of local events. Chemicals are soon mixed throughout the oceans and provide a world average. Whereas the direct record of ancient glaciation on land is prone to be worn away and lost, marine strata preserved in even piecemeal sequences and far away from glaciers may well hold chemical documentation.

Sea-level changes. Both the timing and relative strength of ancient glacial waxings and wanings may be revealed by sea-level changes if the changes can be separated from those resulting from tectonic and local sedimentation fluctuations. Sea level is lowered in proportion to the volume of ice contained in glaciers sited upon land that depresses the crust. When the ice melts, the crust rebounds according to reasonably well understood principles of glacial isostasy (Andrews, 1974; Wilgus et al., 1988; NAS-NRC Panel, 1990). During late Cenozoic times ice accumulated nearly simultaneously in the Northern and Southern Hemispheres as well as on mountain ranges in intermediate latitudes (Denton and Hughes, eds., 1981), and it probably did so during ancient glaciations, although not precisely (Blunier et al., 1998). The architecture of facies arrangements in Cenozoic glacial regions may disclose changes in sea level (Boulton, 1990).

Measurable changes of sea level during the waxings and wanings of Pleistocene glaciers took place between 10 ka and 750 ka

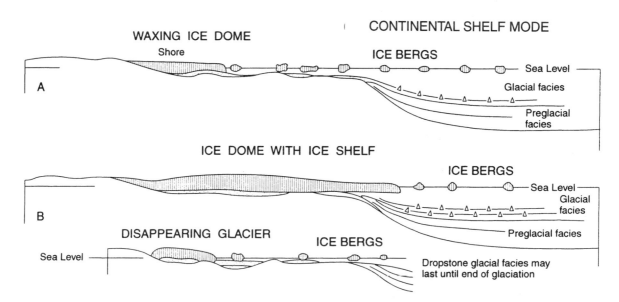

Figure 24. Facies relations associated with growth, culmination, and dwindling of continental glaciers: Continental shelf mode. Glaciers grow in shallow water at and near sea level. Icebergs may spawn early in the glacier's history, and the iceberg phase may last nearly throughout its life. The preglacial beds below the glacigene layers are only slightly older than those documenting glaciation. (See text and compare with Figure 23.) From Crowell (1983a, fig. 1).

(Hays et al., 1976; Imbrie and Imbrie 1980; Berger et al., 1984; Crowley and North, 1991). These variations are now related to changes in the orbital arrangement of the Earth as it moves about the sun. Similar fluctuations are recognized in parts of the pre-Cenozoic glacial record, and similar complexly interconnected causes are likely (Fischer, 1982; Fischer and Schwarzacher, 1984).

Sea-level changes caused by processes other than those due to the waxing and waning of glaciers must be discriminated. These processes are those that change both the volume of water available to fill ocean basins and the shape of the sea and ocean floor. Under the premise that the volume of water on Earth either has been nearly constant throughout geologic time or has changed but slowly and regularly, the amounts drawn into ice caps remove water from the oceans and lower sea level. Calculations on the drop in sea level for the Wisconsin-Weichselian Ice Stage of the Quaternary, including allowance for isostatic loading, give a range between 127 m and 163 m depending on whether a minimum or a maximum reconstruction of ice sheets is selected (Denton and Hughes, eds., 1981). This range fits well with local observations of sea level changes and from oxygen isotope calculations (Fairbanks and Matthews, 1978). During the maximum spread of ice sheets upon Gondwana during the Carboniferous, an estimate of the isostasy-corrected glacio-eustatic component of sea level change is 60 ± 15 m relative to the present (Crowley and Baum, 1991a), and model studies suggest seasonal cycles (Crowley et al., 1987). Investigations of sea-level variations in bringing about the Pennsylvanian cyclothems in North America and Europe (Heckel, 1986; Ross and Ross, 1988; Boardman and Heckel, 1989, 1992; Van Veen and Simonsen, 1991; Maynard and Leeder, 1992) and in southern Africa (Visser, 1993a) suggest that ice waxing and waning at that time have accounted for ranges in sea-level changes between 100 m and 200 m. Soreghan and Giles (1999) advocate eustatic amplitudes in excess of 100 m during Pennsylvania time. Harrison (1990), in calculating changes since 80 Ma—the time of Late Cretaceous continental flooding—gives a maximum range of ~230 m. Ancient ice ages, viewed as of similar or somewhat greater severity than those of the Pleistocene, can therefore reasonably account for sea-level drops of between 100 m and 230 m, perhaps as much as 250 m. Such a lowering and subsequent rising should be discernible in the geologic record.

If huge synchronous ice sheets are advocated for the remote past, such as those requiring a lowering of very much more than 230 m, this lowering should be detectable. Eustatic changes greater than those recognized in the late Cenozoic have not been documented for ancient ice ages except perhaps in the Neoproterozoic of Australia, but here they may be associated with quite local tectonic mobility where submarine canyons cut into a flexed rift margin (Eickhoff et al., 1988; Christie-Blick et al., 1990b; Young and Gostin, 1990). In contrast, the Neoproterozoic "snowball Earth" hypothesis, as described by Kirschvink (1992) and Hoffman et al. (1998a,b), perhaps involved only very thin continental glaciers with little water tied up within them. If so, the effect on sea level would be relatively small.

Tectonic sea-level changes are local, regional, and even worldwide in their effects. Those related to rates of sea-floor spreading and the positioning and size of mid-ocean ridges modify the shape and volume of major ocean basins (Sclater and Francheteau, 1970; Hays and Pitman, 1973; Sclater et al., 1977; Pitman, 1978; Christie-Blick et al., 1990a; Harrison, 1990). During times of rapid spreading, the mid-ocean ridges are broad and high and displace oceanic water upon continental shelves. Ridge flanks are relatively warm and therefore expand upward. When spreading rates are slow, ridges and their flanks stand lower because they cool and contract, and sea level falls. These fundamental changes are rooted in the internal thermodynamics of the Earth and largely control continental elevation and freeboard and in turn influence the likelihood of glaciation and the rate of biotic changes. Variations in the pace and pattern and vigor of plate-tectonic activity need evaluation when appraising the global level of the sea and its effect on climate change.

Long-term changes in sea level are commonly referred to as first-order changes, and contrast with those of lesser magnitude and greater frequency such as those disclosed on plots of sea level through time (Vail et al., 1977; Haq et al., 1988; see also Miall, 1992, for a contrary view). More frequent variations are referred to as second-, third-, and fourth-order departures. These plots, first published by Exxon geologists with only sparse published supporting data, are based on interpretations of seismic profiles and dating of stratal prisms lying at continental margins. Recently, following upon two decades of discussion and controversy, details of the "Exxon sea-level curve" have received considerable confirmation with a sea-level variation over ca. 1 m.y. of as much as 140 m, for the middle and upper Eocene with a variation above the detectable limit of ~20 m, and for the Early Cretaceous with a variation of ~100 m (Miller et al., 1996b; Pekar and Miller, 1996; Stoll and Schrag, 1996; Browning et al., 1996).

Some of these variations are probably the result of fluctuations in glacier volumes so that waxings and wanings over short time intervals are implied. Other factors influencing short-term sea-level variations that require evaluation include variations of tectonically induced stress within tectonic plates (Cloetingh, 1986, 1988). Thermomechanical modeling suggests that tectonically caused vertical motions of the crust in combination with fluctuations in sedimentary loading may produce apparent sea-level changes of as much as 50 m. In addition, the following processes also need appraisal: variations in sediment flux, volcanism beneath oceans, sedimentation of calcium carbonate and terrigenous debris that displaces ocean water, and ocean expansion or contraction due to climatic warming or cooling. It is indeed difficult to disentangle the effects of glacier growth and melting from the other influences causing a rise and fall of sea level.

Cycles, cyclothems, rhythmites, and varves. Bedded sequences, especially those of thin-bedded or laminated strata, at many places and of many ages reveal cyclic repetitions in lithologic types with differing thicknesses (Duff et al., 1967; Einsele and Seilacher, 1982; Kemp, 1996). It has long been recognized, for example, that vertically repeated sequences of facies in cratonic regions (cyclothems) are the result of alternating transgressions and regressions of sea level, and these might be caused by the waxing and waning of far-distant glaciers (Wanless and Shepard, 1936; Crowell, 1978; Fiege, 1978; Heckel, 1986, 1995; Boardman and Heckel, 1989; Miller et al., 1996a, b). Cyclothems especially are described in detail from Pennsylvanian and Permian strata in the midcontinent region of the United States, and from upper Carboniferous beds of similar age in Europe (Maynard and Leeder, 1992). Similar cycles have also been recognized in the western United States (Dickinson et al., 1994; Houck, 1997). Typical cyclothems usually begin with a disconformity at the base, overlain by nonmarine sandstone, at places pebbly, and overlain in turn upward successively by sandy shale, freshwater limestone, underclay, coal, shale, marine limestone, laminated shale, and marine limestone. The sequence is explained as the result of a transgressive sea moving across a stable craton with a gentle seaward slope, with the shoreline represented by the passage of the gradation between coal and marine shale. Upon regression, some erosion takes place, but in the main the sequence is reversed and preserved. More than 100 of these repetitions are recognized and have been followed laterally for several hundreds of kilometers (Moore, 1964; Ross and Ross, 1985).

The rise and fall of sea level as recognized by cyclothem facies and other sedimentary cycles is influenced by irregularities in sediment influx and depositional patterns, local tectonics, glacio-eustasy and by orbital variations. As a consequence, it is difficult to sort out the part of the rise and fall owing to distant glaciation from these other influences. Moreover, the timing of the sea-level changes is at present largely unknown, a timing needed to place bounds on rates of these changes (Klein and Willard, 1989; Klein, 1990, 1992; Langenheim, 1991; Klein and Kupperman, 1992). For example, there is no assurance that the variations fit into recognized orbital cycles, isostatic ups and downs, or the timing of possible stress-induced cycles. Nonetheless, when in the future these influences are adequately appraised, a significant glacial control on cyclothems is expected. In fact, the timing of waxing and waning of distant glaciers is more likely to be determined by investigations of cyclothem timing than from the ephemeral stratal record near Gondwanan glaciers that thrived at the same time. Current understanding suggests that the glaciers waxed and waned with frequent repetitions, but the cyclothems cannot reveal where the waxings and wanings took place. They provide a complex but integrated record of sea-level ups and downs, a large part of which is deemed as rooted in changes in the total mass of glaciers as ice centers moved across Gondwana during its migration across the South Pole. It is notable that cyclothems are well documented for times corresponding to the late Paleozoic glaciations but not for older ice ages. Perhaps this older record has either been eroded or not yet recognized and investigated. Searches for pre Carboniferous cyclothems are needed.

Laminations deposited annually, especially in periglacial lakes, are known as *varves* (Mörner, 1978). They form where lakes are frozen over in the winter so that bottom sedimentation is inhibited, followed by open-water sedimentation after the spring breakup. In Swiss mountain lakes the varves are characterized by summer calcite precipitation when biologic activity thrives, followed in the winter by weak detrital sedimentation (Hsü, 1985). Where the annual cyclicity cannot be demonstrated adequately, laminated sequences are termed *rhythmites*. Since about 1884 correlation of recognizable varves from Quaternary lake to lake in Scandinavia has led to the erection of a chronology for the past 12 k.y. or so (De Geer, 1912) under the premise that the varves are truly annual. Processes other than annual overturns may cause thin laminations, however, such as weak distal turbidity currents, tidal currents, settling of wind-blown detritus, or settling from plumes discharging cloudy water upon the surface or into intermediate layers of a lake, sea, or ocean. Varve-like strata are recognized for most of the ancient ice ages described here, usually from local outcrops or quarries, but as yet not enough careful petrologic work has been undertaken to prove that they are indeed annual varves.

Stratal record of extraterrestrial influences

Bolide impacts, systematic changes in the Earth's orbit, tilt, or wobble as it circles the sun, variations in solar heat flux, and modifications in the amount of space dust shielding the Earth are all viewed as affecting the heat arriving at the Earth's surface. To recognize or document these effects on climate within the ancient rock record, however, is daunting and burdened with difficulties. Theoretical and model studies guide geologists in their search for evidence of bolide impacts and orbital and other variations, and progress is underway (House and Gale, 1995). It is likely that future investigations will disclose times when these and other extraterrestrial events strongly influenced climate.

Bolide impacts. Where asteroids, meteorites, or comets impact the Earth, craters and deposits of jumbled blocks embedded in diamictite may be strewn out in the vicinity of the crater, either near at hand or at distance. The bolide origin, however, can be surely identified if a crater is found and properly identified on the basis of local and regional structure showing disrupted rocks, and on the occurrence of shatter cones and characteristic fracturing (Gostin et al., 1986; Williams, 1986b; Oberbeck et al., 1993; Rampino, 1994). In addition, the jumbled material thrown out near the crater may contain telltale clues. In Miocene rocks of Germany (Horz et al., 1977, 1983) and in Eocene-Oligocene strata of Spain (Ernstson and Claudin, 1990), shatter cones and nearby impact craters of appropriate age confirm a bolide origin. A huge bolide that struck the Earth at the end of the Cretaceous is focusing investigations on the Chicxulub crater, near the Yucatan Peninsula, Mexico, inasmuch as it is suspected to be the culprit scar (e.g., Alvarez, L. W., et al., 1980; Alvarez, W., et al., 1995; Alvarez, W., 1997; Morgan et al., 1997). Older impacts are reported within the geologic record, and some of these impact events are dealt with briefly in the discussion of ice ages below, such as those near the Devonian-Carboniferous transition. The Neoproterozoic Acraman impact structure of South Australia, ca. 600 m.y. old, has been well documented along with debris thrown from it for at least 350 km eastward and 450 km northward of the source region of the impact crater (Gostin et al., 1986; Williams, 1986b; Schmidt and Williams, 1991). Other impacts are recorded in Australia but with large age uncertainties (Shoemaker et al., 1990). Sand-sized spherules resulting from Archean bolide impacts have even been recovered from South African strata ca. 3,400 m.y. old (Lowe et al., 1989). As yet, however, there is very little in the rock record to indicate that impacts arrived with rhythmic timing (Rampino and Stothers, 1984; Shoemaker and Wolfe, 1986). The worldwide record at hand includes only about 150 impact craters, but the search continues for both the craters themselves and other bolide spoors (Grieve et al., 1995; McKinnon, 1997). Evidence of such events through time, including impactites or ejectites (and tektites and biogeochemical modifications), indicates that other bolides may have been overlooked or misidentified, or remain undiscovered. If a bolide origin is suspected, careful study of the strata and any product that might have been thrown out during an impact is required. Many more impact events probably remain to be discovered. Craters of many are lost, however, such as those from bolides that fell into pre-Jurassic oceans, and the sea floor has been lost by crustal subduction.

Orbital cycles. As the Earth moves about the sun through geologic time, it has acquired differing elipticities and tilts and wobbles as the result of orbital or Milankovitch cycles. These bring about systematic changes in the solar heat flux arriving at the Earth's surface that, in turn, affect the climate and show up through the waxings and wanings of glaciers along with several other effects. Much of the research on orbital effects has concentrated on the Quaternary, when and where measurable changes of sea level can be correlated with the growth and decay of glaciers. These cycles took place over time periods between 10 ka and 750 ka which fit Milankovitch cycles quite satisfactorily (Hays et al., 1976; Imbrie and Imbrie, 1980; Kukla et al., 1981; Berger et al., 1984; Crowley and North, 1991; Sanders, 1995).

Similar variations are now recognized for pre-Pleistocene times with similar cycles. For example, Pliocene turbidites in Greece, laid down in the Mediterranean Sea when it was not open to the world ocean and not accessible to worldwide sea-level changes related to distant glaciation, are reported to disclose an orbitally controlled cyclicity (Weltje and de Boer, 1993). Probable orbitally paced climate oscillations occur across the Oligocene-Miocene boundary, at a time when the Late Cenozoic Ice Age was beginning to get underway (Zachos ct al., 1997). Some of the cyclicity identified in Cretaceous strata is interpreted as related to orbital forcing at a time of minimal glaciation (Fischer, 1982, 1986; Fischer and Schwarzacher, 1984). Jurassic microrhythmites have been identified in Britain, again interpretable as resulting from orbital variations (House, 1985a), but Triassic rhythmites in the southern Italian Alps probably do not fit (Brack et al., 1996). Triassic laminites in the Newark basin of the eastern United States, on the other hand, appear to match orbital time scales (Olsen, 1986; Olsen and Kent, 1996). Although some cyclicity reported through the geologic column may be related to glacier

growth and decay under the influence of orbital cycles, many other factors also need appraisal. Theoretical and model studies are becoming helpful in recognizing these relationships and guiding geological field searches (Crowley et al., 1992). Unfortunately, as well, for most ancient glacial sections, dating is imprecisely known. Only under unusual circumstances, such as when datable ashbeds are present, is the time span of deposition known. For ancient strata seldom are there acceptable data leading to interpretations of rates of deposition, or the timing involved in any recognized cyclicity.

A useful undertaking to find out whether laminations in ancient stratal sequences are indeed cyclic would be to study carefully a very well exposed stratal section in three dimensions, where stacked laminations can be followed laterally and confidently for several hundred meters. Such a section in ancient strata, very well exposed in a quarry or roadcuts, might contain distinctive marker beds that can be traced with confidence throughout the area of exposure. The number of laminations between two such marker beds, with several meters of laminations between them, must be the same throughout the exposures if there is cyclicity. If not the same, it is likely that weak currents have laid down laminations that pinch out laterally. Only if cyclicity can be established with satisfactory time resolution can repetitions due to orbital variations, tidal-current influences, solar variations, or other processes be separated. Such studies would substantiate investigations of long cores or single stratigraphic sections employing time-series analysis.

Other extraterrestrial influences. There are other extraterrestrial influences affecting Earth's climate, and so presumably the pacing and strengths of glaciations. These have been widely discussed, but the record within rocks or revealed by biogeochemical tracers is sparse. They include interplanetary or cosmic dust coming between the sun and Earth (Hansen and Lacis, 1990; Shoemaker et al., 1990; Taylor, 1992, 1998; Anderson, 1993; Brownlee, 1995; Farley, 1995; Farley and Patterson, 1995; Hughes, 1996; Taylor et al., 1996). The sun's heat output varies through time and astronomic theory indicates that it may have been increasing in luminosity since the Earth's origin (NAS-NRC Panel, 1982; Eddy et al., 1982; Evans, 1982; Radick et al., 1990; Kelly and Wigley, 1992; Kasting, 1993). In addition to the cyclic variations in the eccentricity of orbits and orderly changes in the tilt of the rotational axis to the orbital plane, the inclination of the orbital plane itself may have varied through time and may be identifiable in ocean-floor cores (Muller and MacDonald, 1995, 1997; Kortenkamp and Dermott, 1998). The origin and arrangement through time of the Earth-moon system and changes in tides have received attention (e.g., Sonett et al., 1988, 1996). Other effects proposed include galactic causes, including among others the passage of the solar system through a dusty spiral arm of our galaxy (e.g., Steiner and Grillmair, 1973; McCrea, 1975, 1981; Berger et al., 1985; Rampino and Stothers, 1984). It behooves the geologist to find suitable ways to detect these suggested effects within the rock record.

Volcanism

Volcanic eruptions affect climate through the effusion of greenhouse and other gases and aerosols, and by providing local heat sources (e.g., Bryson and Goodman, 1980; Axelrod, 1981; Rampino et al., 1988; Symonds et al., 1988; AGU, 1992). Sulfur dioxide and other gases thrown into the atmosphere by El Chicon (Mexico, 1982) and Mount Pinatubo (Philippine Islands, 1991) affected worldwide climate and even influenced the ozone layer (Arnold et al., 1990; Grant et al., 1992; Wallace and Gerlach, 1994; Minnis et al., 1993; McCormick et al., 1995; Briffa et al., 1998; de Silva and Zielinski, 1998). The huge eruption of Toba in Sumatra ca. 73.5 k.y. ago probably caused several years of cooling by increasing the albedo within the upper atmosphere (Rampino and Self, 1992). Pleistocene volcanism is even argued as being involved in glacial initiation (Kennett and Thunell, 1975; Bray, 1977). Iron from volcanic ashes may increase the amount of atmospheric oxygen by increasing the breakdown of sulfate in marine sediments as has been suggested for lower Paleozoic rocks (Spirakis, 1989). Mantle plumes are deemed to have been associated with flood basalts and climate change in the Ethiopean region at least as far back as 30 m.y. ago (Hofmann et al., 1997). In the more remote past marked arc volcanism, especially near the tropospheric tropical convergence, must have thrown gases and aerosols high into the atmosphere to influence climate. Such correlations have not yet been established, however, because reconstructions of plate arrangements in pre-Mesozoic time are not satisfactory enough to show where island and continental volcanic arcs were located with regard to air-ocean circulations. Research is also focused on the timing of volcanic outpourings, especially plateau basalts, with respect to events in the evolution of life and major rearrangements of tectonic plates.

Geochemical cycles, greenhouse gases, and the glaciation record

Several geochemical cycles and their changes in flux throughout geologic time bear on understanding climate. Especially important are variations in the "greenhouse" gases that modulate the heat retained at the Earth's surface from the sun's radiation. The contribution of heat from the Earth's interior is proportionately so much less than that from the sun that it can be ignored (Verhoogen et al., 1970). Throughout time, changing configurations of land and sea and locations of volcanoes spewing out gases and aerosols have especially affected the distribution of these gases as well as changing fluxes of dust and aerosols. Both biologic and physical processes have modified contributions to the air-ocean system as life has evolved. Fortunately, pathways of some of these contributions are revealed and traced by isotopic studies, such as those of carbon, oxygen, sulfur, and strontium preserved within sediments and in fossils (Fig. 25). Because of their rapid mixing throughout the world ocean, geochemical measurements may display global distributions and therefore provide useful guides to global paleoclimate.

The greenhouse gases include water vapor, carbon dioxide, and methane, along with some trace gases such as nitrous oxide (e.g., Peixoto and Oort, 1992; Holland and Petersen, 1995). They are quite minor constituents of the atmosphere, which consists mainly of ~78% N_2, ~21% O_2, and ~1% Ar, but very significant climatologically although they make up only ~0.04% by volume. Water vapor varies in concentration through a large range from ~40 parts per million by volume (ppmv) to ~40,000 ppmv depending directly on the temperature. Carbon dioxide concentrations range around 360 ppmv, methane ~1.7 ppmv, N_2O 330 parts per billion by volume (ppbv), and NO_2 and NO each ~1 ppbv. These gases are strong absorbers of long wave radiation, so that heat arriving from the sun is trapped in the lower atmosphere by them and the temperature raised. With an increase in temperature the amount of heat radiated back to space is also increased by reflection (the albedo) from clouds, resulting in a "negative feedback" and a temperature lowering. The heat-blanketing effect of lower clouds retains heat in the lower atmosphere, tends to raise the temperature, radiates some heat downward toward and into the Earth's surface. Clouds, composed of water condensed upon nucleii consisting of minute particles such as dust, tiny ice crystals, and aerosols, reveal the structure of the atmosphere where water vapor is high although water vapor itself is invisible.

Even today, however, understanding the complexities of this atmospheric heat balance system is quite incomplete. Necessary research is concentrating on measuring and evaluating albedo and other factors in the system (e.g., Slingo, 1990; Kiehl, 1994; Baker, 1997). The role of aerosols is puzzling, such as the aerosols coming from oceanic phytoplankton contributing dimethylsulphide to serve as cloud-condensation nucleii, particles of sea salt, and many others (e.g., Charlson et al., 1987; Taylor and Penner, 1994; Murphy et al., 1998). Humankind's contributions to aerosols and carbon dioxide and their influence on climate is receiving attention especially (e.g., Falkowski et al., 1992; Cicerone, 1994; Andreae, 1996; Seinfeld, 1998). Inasmuch as the interplays of these many processes and products are poorly understood today, we can speculate only cautiously on their influences on climate in the geologic past.

Some concepts dealing with the biogeochemical influences on climate, especially as revealed by isotopic tracers, are reviewed briefly here but are dealt with more satisfactorily in many references (e.g., Holland, 1978, 1984; Savin, 1982; Schidlowski et al., 1983; Hoefs, 1987; Gregor et al., 1988; Kump, 1989; Berner, 1987, 1990, 1991, 1993; Schidlowski, 1988; Wilkinson and Walker, 1989; Pederson and Calvert, 1990; Berner and Caldeira, 1997; Bickle, 1998; Caldeira and Berner, 1998). Chemical weathering of rocks on land is subdued during times of coolness, as in the vicinity of glaciers, and physical weathering processes are proportionately enhanced (Young, 1969; Nesbitt and Young, 1982, 1996; Nesbitt et al., 1996; Anderson, 1997; Panahi and Young, 1997; Young et al., 1998; Young and Nesbitt, 1999). Products of these processes may therefore reveal information on the relative importance of these processes and in turn on climate.

Materials within tillites, both stones and matrix, will display evidences of strong physical disintegration and little evidence of chemical alteration if both clasts and matrix were derived directly from cratonic basement terranes and not from reworked marine sediments that are the product of an older weathering and depositional cycle. The bulk composition of diamictite matrix and interbedded mudrocks will be expected to be close to that of cratonic basement if the source region was undergoing glaciation. If chemical alterations predominate, derivatives from feldspars in the basement source area, for example, will increase in abundance. A Chemical Index of Alteration (CIA) has been introduced to quantify such measurements, referred to in discussions below (Nesbitt and Young, 1982; Nesbitt et al., 1996; Fedo et al., 1995; Young and Nesbitt, 1999). High CIA values are related to the removal of

unstable (labile) cations (e.g., Ca^{2+}, Na^+, K^+) relative to stable residual constituents (Al^{3+}, Ti^{4+}); low values reflect cool weathering in the source region.

Carbon dioxide and methane became influential upon climate very early in Earth history and especially as life evolved (e.g., Cloud, 1968, 1976, 1988; Walker, 1977; Holland, 1984; Schopf and Klein, 1992). A decrease in carbon dioxide favors coolness and glaciation (Tans et al., 1990; Raymo, 1997). Both carbon dioxide and methane receive much attention today because the quantities being added to the system are related to human activities as the result of burning fossil fuels, deforestation, and husbandry of flatulent ungulates such as cattle and sheep. Changes in climate owing to human activity are geologically ephemeral. In taking a long-time view of Earth, they repre-

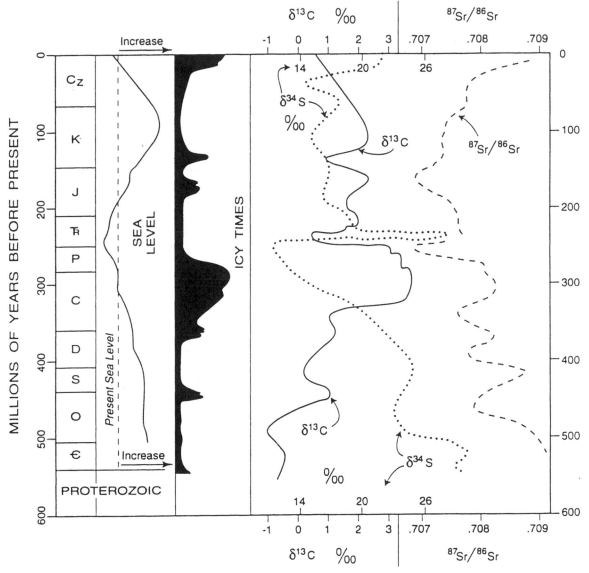

Figure 25. Carbon ($\delta^{13}C$), strontium ($^{87}Sr/^{86}Sr$), and sulfur ($\delta^{34}S$) isotopic ratios through Phanerozoic time compared with sea level and icy times. Cz = Cenozoic; K = Cretaceous; J = Jurassic; Ŧ = Triassic; P = Permian; C = Carboniferous; D = Devonian; S = Silurian; O = Ordovician; Є = Cambrian. Modified from Veevers (1994a, fig. 1, compiled from several sources).

sent but a quick excursion, presumably one that will return to a normal balance as humankind evolves into whatever form of life succeeds us.

At warm times in the past greenhouse gases have been more abundant, giving rise to the concept of a "Greenhouse Earth," In cool times, greenhouse gases have been less abundant and glaciers have thrived in an "Icehouse Earth" (Fischer and Arthur, 1977; Arthur, 1982; Fischer, 1982, 1986). The terms are useful and widely used, but perhaps overemphasize the role of greenhouse gases in the complex system. Moreover, their use is widely associated with the concept of cyclic time repetitions, but, as shown on Figure 1, there is little evidence of cycles between warm and cool intervals if nearly equal durations between intervals are implied. A regular cyclic repetition of glacial times is not noticeable. I prefer to use descriptive terms such as *icy* or *cool times* and *warm times* or *intervals*, with no emphasis on greenhouse gases although their relative proportion is indeed important. Other factors are also significant.

Water vapor, clouds, and precipitation belts. Water vapor is a very significant and the most variable greenhouse gas, especially on regional scales. Its influence on clouds, albedo, and the locations of precipitation belts through time as the continents and seas have become rearranged has not yet been adequately examined. The climatic influence of water vapor and cloud and precipitation belts is illustrated by the situation in the North Atlantic region today. Because of the shape of the North Atlantic Ocean with respect to the flow directions of air and ocean currents, warm surface water reaches into very high latitudes, so that even Murmansk on the northern coast of Russia usually remains ice free during winters. Many locations at this high latitude of ~69° N are icebound in winter. Before the rising of the Panama Isthmus owing to tectonic and volcanic processes in the late Miocene and Pliocene, surface waters from the Atlantic flowed westward out of the Atlantic and into the Pacific, and were not then deflected to the far north by ocean currents, including the Gulf Stream (Crowell and Frakes, 1970; Ross, 1996; Haug and Tiedemann, 1998; Driscoll and Haug, 1998). Warm waters evaporate much more easily than cold waters and feed precipitation as snow on nearby land—snow that lasts from year to year and grows into glaciers. During the Late Cenozoic Ice Age, this chain of circumstances enhanced the growth of glacier centers in Scandinavian regions and Greenland. Glaciers along the uplifted margins of the North Atlantic region (Eyles, 1996), along with associated bathymetry and topography, not only have strongly affected the local and regional climate, but also have influenced the climate of the Earth as a whole. For example, the pattern of heat transport in the oceans today is viewed as controlled by the thermohaline "conveyor-belt" system, which in turn is controlled by the arrangement of continents and their topography and the oceans and their bathymetry (Broecker et al., 1985; Broecker and Denton, 1989; Broecker, 1997).

As another example, current continental arrangement and topography bring about the Indian monsoon, which determines the climate for large parts of southern Asia. The rising of the high-standing Himalayan and Tibetan regions, with the warm Indian Ocean lying to the south in tropical regions, drives the monsoon today. This topographic and monsoon arrangement is also viewed as largely responsible for the onset and growth of the Late Cenozoic Ice Age (Molnar and England, 1990; Raymo and Ruddiman, 1992; Raymo, 1994a,b). Similar arrangements and processes are deemed to have operated in the past, arrangements that affect not only large regions but also the climate of the Earth as a whole.

How often in the past have similar changes in land and sea arrangements brought warm water to high latitudes to precipitate as snow that later consolidated into ice or, in contrast, kept warm water in near-equatorial regions so that it did not flow poleward? How often have continental arrangements favored longitudinal instead of latitudinal oceanic circulation? What have been the bathymetries of ocean floors to affect the flow of bottom waters and the thermohaline heat exchange between high and low latitudes? How have "conveyor-belt" systems operated in the past? For pre-Mesozoic times, however, there has been but little consideration of the functioning of such processes inasmuch as paleogeographic reconstructions are quite crude although there have been a few speculations (Crowell, 1978). In summary, I view the tectonic arrangements of the continents within the world ocean as of paramount significance in influencing both regional and world climate in the geologic past through their effect on water vapor, cloud, and precipitation-belt distributions. They are probably as important as changes in the amount and distribution of other greenhouse gases, including carbon dioxide.

Carbon. The long-term carbon cycle involves the exchange of carbon, especially carbon dioxide, between rocks and the combined atmosphere, ocean, biosphere, soil, and deep-earth system including volcanism. The system includes the weathering of silicates and organic matter on continents, burial and later exhumation of organic matter and carbonates in sediments, and the breakdown of organic matter and carbonates as they are heated at depth. These processes tend to remain approximately in balance except during unusual events such as those near the end of the Proterozoic, during episodes of unusual widespread volcanism, and for a time after bolide impacts. When carbon dioxide amounts have increased significantly, feedback mechanisms have prevented the increase from getting out of hand (Berner and Caldeira, 1997; Bickle, 1998; Caldeira and Berner, 1998). An explanation is probably that with more carbon dioxide in the atmosphere the temperature is increased along with rainfall, which in turn enhances its removal through faster silicate weathering. The complicated carbon system through time has attracted much attention but requires more investigation before we truly understand its influence on climate (Fischer, 1982, 1986; Arthur, 1982; Holser et al., 1988; Worsley and Nance, 1989; Raymo, 1997). The geological significance of the carbon cycle was probably first recognized by Arvid Gustaf Högben in 1894 (referred to in Berner, 1995; Fleming, 1998) and then by Svante August Arrhenius in 1906 following upon discussion of the greenhouse effect (referred to in Fleming, 1992, 1998; Weart,

1997). Schermerhorn (1983) raised the likelihood of CO_2 depletion in bringing about the Late Paleozoic Ice Age.

The circulation of carbon and its isotopes through the Earth system is primarily controlled by biogenic activity in combination with the pattern of oceanic currents, tectonic processes and styles, volcanism, weathering and erosion on land, and sedimentation and carbon burial within sediments and its later release into the sea (Arthur, 1982; Walker et al., 1983; NAS-NRC Panel, 1983; Sundquist and Broecker, 1985; Faure, 1986; Holser et al., 1988; Wilkinson and Walker, 1989; Worsley and Nance, 1989; Tans et al., 1990; Crowley and North, 1991; Sundquist, 1993; Ruddiman, 1997). Fortunately, isotopes trace some carbon pathways. Isotopic fractionation is mainly the result of photosynthesis where the heavier isotope (^{13}C) is separated from the lighter (^{12}C) that remains in the atmosphere. The proportion is noted by the ratio $^{13}C/^{12}C$, usually given as $\delta^{13}C$, the departure with reference to a conventional standard (the Peedee belemnite). Excursions to more positive values of this ratio are usually interpreted as indicating that more organic carbon is being buried and sequestered, removing it temporarily from the global pool. High minus values suggest sharp reduction in biologic activity, especially a reduction in photosynthesis.

Increasing elevations within orogenic belts or within continents increase silicate weathering rates and decrease carbon dioxide amounts, although increased sedimentation in nearby basins may operate to bury carbon and remove it from the system temporarily (France-Lanord and Derry, 1997). In addition, the flux of carbon dioxide from deep degassing and metamorphism and hydrothermal activity—all consequences of plate-tectonic vigor and pattern—also affects the complex system. Deformation and uplift of carbon-containing sediments within orogenic belts may release sequestered isotopically light organic carbon into the Earth's fluid system and shift the ratio toward lighter isotopic values (Beck et al., 1995). As an example of the usefulness of isotopic tracers, the ratios of $^{12}C/^{13}C$ successfully track temperature, is revealed by associated $^{18}O/^{16}O$ ratios from late Cenozoic ice cores from Antarctica and Greenland (e.g., Barnola et al., 1987; Curry and Crowley, 1987; Raymo et al., 1989; Worsley and Nance, 1989; Lorius et al., 1990; Crowley and North, 1991).

In studying ancient climates more understanding of the carbon system is needed to answer such questions as: How do we evaluate the alternate removal and release of carbon to the air-ocean system as sea level rose and fell during our appraisal of paleogeographic and paleotectonic reconstructions? Were there broad and shallow continental shelves to retain carbon in marine sediments when sea level rose that then became available to weathering as sea level fell? Were there high mountain ranges, such as the Himalayan chains that rose in Cenozoic time, to influence the balance between physical and chemical weathering and therefore the ratio of carbon dioxide to oxygen (Raymo and Ruddiman, 1992; Raymo, 1994a,b, 1997)?

Throughout geologic time and beginning well before the first known glaciation, the ratio of carbon to oxygen has been influenced by biogenic activities (Schidlowski, 1988; Schopf,

1993). There is a relatively close correlation in the Phanerozoic between low carbon dioxide abundances and the occurrences of ice ages, as discussed below (Hudson and Anderson, 1989; Berner, 1990, 1992, 1997). Marked isotopic excursions of carbon occur during Late Permian and Triassic intervals when sea level was low so that isotopically light buried carbon within sediments was weathered and released. It was recirculated into the oceans and atmosphere, increasing carbon dioxide proportionately. Flood volcanism, especially the outflow from Siberian traps and ignimbrites along the Panthalassan margin of Gondwana at these times increased both the discharge of heat and carbon dioxide (Renne and Basu, 1991; Veevers et al., 1994d). In the Late Proterozoic, carbon isotopic excursions are especially marked, pertinent to the concept of a "snowball Earth" as discussed below (Harland, 1964; Kirschvink, 1992; Knoll and Walter, 1992, 1995; Ripperdan, 1994; Hoffman et al., 1998a,b; Kennedy et al., 1998). However, over much of geologic time the $^{13}C/^{12}C$ ratio of dissolved inorganic carbon in sea water has remained close to its present-day value implying that the proportion of carbon in the active organic and inorganic reservoir has remained relatively constant (Schidlowski, 1988). Changes in the carbon dioxide partial pressure in the past have been recognized in Cenozoic and Mesozoic soils (Cerling et al., 1989) and in Ordovician goethites in ironstones (Yapp and Poths, 1992). Understanding the complicated carbon cycles is still very much needed; this understanding purports to reveal much concerning ancient climate history.

Oxygen. Oxygen entered the system first in chemical combination with other elements and then as free oxygen gas beginning in mid-Proterozoic time and then accelerating during the Phanerozoic (Cloud, 1988). Changing ratios of light to heavy isotopes of oxygen may reveal information on the waxing and waning of ice on land, especially for the Cenozoic record. Fresh water and glacial ice and snow are enriched in ^{16}O (the light isotope) with respect to ocean water, which is enriched in ^{18}O (the heavy isotope). This difference in ratio between the light and heavy isotopes of oxygen is detectable in the calcium carbonate of marine foram tests and other fossils and deposits (Urey, 1947; Emiliani, 1955; Dansgaard, 1964; Dansgaard and Tauber, 1969; Shackleton and Opdyke, 1973; Douglas and Woodruff, 1981; Moore et al., 1982; Miller et al., 1987; Shackleton, 1987a, b; Broecker, 1989; Robinson, 1990; Calkin, 1995). The method depends on the circumstance that when and where calcium carbonate is deposited in isotopic equilibrium with open-sea water, the ratio of $^{18}O/^{16}O$ is a function of temperature (Savin, 1982; Faure, 1986; Hoefs, 1987; Hudson, 1989; Hudson and Anderson, 1989). The temperature of carbonate deposition can be calculated if the ambient ratio $^{18}O/^{16}O$ in open-sea water can be satisfactorily estimated and if the carbonate has not been altered since deposition. Inasmuch as the open oceans are well mixed, their isotopic composition is viewed as nearly constant, at least well back into Paleozoic time. It may be difficult, however, to establish from paleogeographic reconstructions that sample sites had access to the open ocean. It is also essential to establish through careful petrographic investigations that detrimental alteration, such as diagenesis, has not taken place. Estimates of temperatures within the carbon cycle depend on comparisons with oxygen isotopic temperatures.

This calcium carbonate paleothermometer is useful back into Mesozoic time, and less so for more ancient strata. Planktonic foraminifera apparently manufacture their tests in isotopic equilibrium with ambient sea water, and benthonic foraminifera less so. Using data from planktonic foraminifera, it is inferred, for example, that Antarctic ice sheets had grown sufficiently large by late-middle Eocene time (at ca. 43 Ma) to also register a glacial-eustatic drop in sea level of more than 20 m (Browning et al., 1996). A decrease of as much as 12 °C is inferred for high latitude surface water using oxygen isotopes from planktonic foraminifera over the past 70 m.y. (Holser et al., 1988; Shackleton, 1987a,b). For ancient strata and their contained fossils it is difficult to establish that significant alterations have not occurred. Even with careful analysis of large amounts of data for ancient carbonate deposits and with a satisfactory determination of isotopic temperature, it is difficult to relate a temperature obtained from marine fossils and sediments to global climate.

Oxygen isotopic ratios obtained from benthonic biogenic silica and phosphates may also give information on temperatures, for example, from cherts deposited in deep water below the calcium carbonate compensation depth (Knauth and Epstein, 1976; Savin, 1982), but there is likely to be large uncertainty of the ambient temperature.

Sulfur. In the sulfur cycle, sulfate is reduced to sulfide and carbon dioxide is released, primarily by the activity of anaerobic bacteria within sediments immediately below the sea floor (Holser et al., 1988; Berner and Petsch, 1998; Paytan et al., 1998). The carbon dioxide in turn, by means of photosynthesis elsewhere eventually yields oxygen so that the processes affect climate mainly by increasing the oxygen proportion. In addition, marine evaporites show variations through time in the $^{34}S/^{32}S$ ratio between the heavy (^{34}S) and light (^{32}S) isotopes. When sea level was low, deposition occurred in isolated basins such as those formed near the end of the Permian. These variations show up as departures or excursions from a conventionally selected standard (primitive values measured within the Canyon Diablo meteorite) and is shown by $\delta^{34}S$ values. In general, the times of maximum deposition of isotopically light sulfur correspond to times when sea level was low and also with times of increased $\delta^{13}C$ in carbonates, such as episodes during Permian and Triassic times. The relation of these excursions to widespread glaciation on land is primarily through their indication of sea-level drop, suggesting emergent continents that are more likely to harbor ice caps than when continents are flooded. In addition, sulfur from volcanic eruptions affects climate through adding sulfate and other products as mentioned above.

Strontium. The isotopic composition of strontium circulated within the hydrosphere depends largely both on the composition of continental rocks subjected to weathering and erosion and on the nature of hydrothermal activity primarily along mid-ocean spreading ridges (e.g., Faure, 1986). In general, cratons composed of old granitic and gneissic metamorphic rocks yield ratios of heavy to light isotopes ($^{87}Sr/^{86}Sr$) that are higher than normal. Mafic rocks under erosion and oceanic sources yield lower values. At present this ratio is nearly constant throughout the ocean so the mixing time for the strontium isotopes is geologically short. In marine carbonates, the ratio has changed measurably throughout geologic time (Peterman et al., 1970; Veizer and Compston, 1974, 1976; Burke et al., 1982). During the past 10 m.y. an increase in the proportion of the heavy strontium isotope corresponding with waxing of late Cenozoic glaciers has been noted (Blum and Erel, 1995). Low values occur in Upper Permian strata, deposited at a time when plate rearrangements were taking place accompanied by vigorous hydrothermal activity emitting the light isotope along mid-ocean ridges. The highest values come from Recent strata, and are perhaps related to the rise and deep weathering in the Himalayan and associated ranges (Edmond, 1992; Raymo and Ruddiman, 1992; Ruddiman, 1997; Raymo, 1994a,b, 1997). On the other hand, the main climatic effect of Neogene Himalayan uplift and erosion may be both a deflection of the jet stream and associated airflow patterns by the latitudinal arrangement of the Alpine-Himalayan ranges and an increase in the amount of carbon sequestered in the huge Bengal subsea fan rather than an increase in silicate weathering fluxes (France-Lanord and Derry, 1997). Plots of the heavy to light strontium ratio show a rise through the Cenozoic but beginning gently in the Jurassic. During this time span this trend suggests a decrease of light strontium from oceanic sources and an increase in heavy strontium from sialic sources. An increase in the ratio during times of Phanerozoic glaciations has been noted (Armstrong, 1971) and during the Huronian glaciation of the Early Proterozoic (Young, 1969). Variable measurements in Silurian beds (Ruppel et al., 1996) and in upper Proterozoic strata are as great as any noted in the Phanerozoic, with low values even lower (Derry et al., 1989; Knoll and Walter, 1992, 1995; Hoffman et al., 1998a,b). These sharp excursions suggest unusual tectonic and climatic conditions near the end of the Proterozoic, as described briefly below. In addition, the excursions are proving to be most useful in interregional correlations.

Other geochemical cycles. The amount of phosphorous preserved as phosphate in a stratal record is linked to rates of continental weathering, the shape and bathymetry of ocean margins and their situation with respect to climate belts and current patterns, high organic activity, and other interrelated factors (McKelvey, 1978; Arthur and Jenkyns, 1981; Sheldon, 1981; Cook and Shergold, 1984; Föllmi, 1995; Van Cappellen and Ingall, 1996; Xiao et al., 1998). Strong oceanic upwelling, where continental margins are oriented longitudinally so that prevailing winds blow consistently across a large ocean with a divergent component to a linear shore, brings cold nutrient-rich waters from depth to stimulate high organic activity. If, upon rise of sea level, shelf areas are inundated and also protected from overwhelming influxes of detrital sediment, phosphates of economic value may form. Such a rise in sea level may result from the melting of continental ice sheets as the climate warms. Near the end of Proterozoic time, for example, valuable deposits were laid down not long after the end of glacial episodes (Worsley et al., 1985, 1986). Permian phosphate beds in basins of western North America constituting the Phosphoria Formation show a similar association.

An increase in iron, perhaps along with manganese, is noted in association with some ancient glacial deposits, such as those of the Neoproterozoic in northwestern Canada (Young, 1976b; Yeo, 1981, 1986). The resulting "banded iron formations" are viewed as resulting from an increase in hydrothermal activity along spreading centers at a time of vigorous plate tectonic activity, including continental fragmentation.

Tectonics of glacial sites and preservation potential of the glacial record

Plate tectonic concepts and patterns are helpful in visualizing both glacier settings and places where glacial debris and facies accumulate (Fig. 26). Rift shoulders are ideal sites for glaciers shedding till intermixed with nonglacial debris that moves downslope to and beyond rift flanks into deep depositional sites on a basin floor (Schermerhorn, 1974; Young, 1976b, 1991; Eyles, 1993, 1996; Deynoux et al. 1994; Link et al., 1994). Rifts enlarging through time to form passive continental margins with buried thick sedimentary prisms may preserve a glacial record. For example, today a widening rift is evolving into a pair of passive margins in the region between Labrador and Greenland where sediment is accumulating in the Baffin, Labrador, and Newfoundland basins (Andrews, 1990; Piper et al., 1990; Eyles, 1993). Another example is the occurrence today of a tectonically active continental margin with an offshore subduction zone along the Gulf of Alaska and the association with characteristic facies, including those derived from glaciers onshore (Eyles and Lagoe, 1990; Eyles et al., 1991; Eyles, 1993). Rifts, forearc, backarc, and deep intracratonic basins may lie within reach of sediment from glaciers and provide receptacles for a record that may survive if these sites are first preserved and then later uplifted and exposed for us to study them.

Sites upon continents where a glacial record is likely to be preserved through very long durations of geologic time are rare. Like the ice itself, which soon melts, the continental record is ephemeral and the skin-deep veneer of continental glacial sediments is quickly worn away. The direct record contained in stratal sequences carrying identifiable glacial facies will be preserved only where deep basins subside and are then tucked down deeper into the crust; a subsequent march of uplifts and erosions to destroy this record has not occurred. The indirect record, consisting largely of geochemical climate proxies, stands a much better chance of coming down to us from ancient times. But even the geochemical proxies, such as isotopic excursions from the normal, become modified by diagenesis, the movement of fluids through rocks, and metamorphism. The strata may remain but important information originally contained within them will have been lost.

When continental masses amalgamate to form supercontinents, heat rising from within the Earth is blanketed so the supercontinents rise and sea level falls and retreats from continental shelves. Exposure of large tracts of continents enhances weathering and the removal of carbon dioxide from the atmosphere and a proportionate increase in oxygen. This, in turn, increases the likelihood of glaciation. With lowered temperatures and an increase in the ice cover, however, weathering is reduced, and a gradual

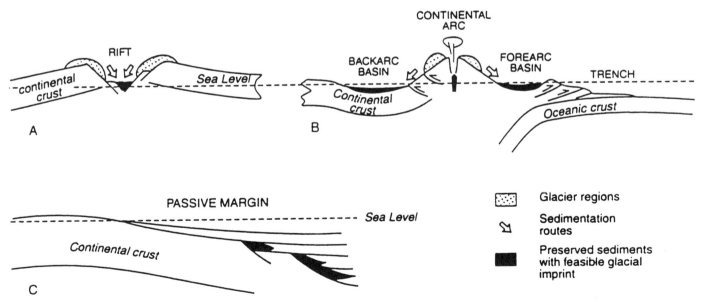

Figure 26. Tectonic styles and their relations to sedimentation sites and likelihood of preservation of the glacial record. (See text.) A = Continental rift setting where glaciers occupy rift shoulders and sedimentary record is preserved within rift sediments; B = continental arc setting where glaciers occur on summits and flanks of volcanoes, and sedimentary glacial record may be preserved within both forearc and backarc basins; C = where a passive margin has evolved from a continental rift, the glacial record may be preserved in deep overstepped layers formed during earlier stages of rifting.

warming ensues from the slow buildup of carbon dioxide. When supercontinents eventually fragment—as the Wilson tectonic cycle progresses—rifting follows with an increase in hydrothermal outpourings, including carbon dioxide. Iron and manganese compounds may become incorporated within "banded iron formation" as during the Proterozoic (Young, 1976b; Yeo, 1981, 1986). The repetition of warming and cooling, along with major plate reorganizations through time, is credited with marked changes in sea level. For example, beginning in mid-Permian time, the huge supercontinent of Pangea began to fragment (Klein, 1994). Sea level rose, encroaching upon continents and reducing their freeboard, and climate warmed.

Tectonic activity seems to have been characterized by long intervals of relatively systematic plate movements interspersed with revolutionary times of rapid movement and reorganization (Gurnis, 1990, 1993; Gurnis and Torsvik, 1994; Kirschvink et al., 1997; Evans, 1998). These major changes in the pace of tectonic activity have in turn affected sea level and climate. Tectonic arrangements may also influence the capacity of ocean basins to hold water (Rona 1973; Veevers and Powell, 1987; Powell and Veevers, 1987; Harrison, 1990). For example, orogenic processes may squeeze continents laterally so that they stand high, bringing about a lowering of sea level. Continents may also stand high because they lie upon parts of the mantle that are less dense than average, as in western North America during the Cenozoic era (Menard, 1973; Crough and Thompson, 1977; Mooney and Braile, 1989; Oldow et al., 1989; Larson, 1991; Anderson, 1994). In fact, the world pattern of types of plate boundaries and intraplate deformations beneath both continents and oceans affects the ocean volume. Most of these tectonic changes are slow and bring about only gradual changes in sea level. Throughout geologic time they are as yet poorly documented.

Major tectonic events bring about complex large depressions such as the Mediterranean basin, which has dried up with a transfer of water to the world ocean and filled up again when reconnected to it. In late Miocene time, such events probably happened to the Mediterranean and Black Sea systems (Hsü et al., 1973) with related changes in eustatic sea level of ~10 m, first up and then down (Berger and Winterer, 1974). Similar events are deemed to have occurred in Middle to Late Jurassic time when an isolated North Atlantic basin originated, grew in volume, and then was flooded (Sclater et al., 1977; Schopf, 1980). This basin was larger than the Mediterranean system today, and so its effect on sea level was potentially greater. The filling of the North Atlantic may account for the sharp regression during the Callovian stage at the end of the Jurassic (Schopf, 1980). Smaller basins, such as those along the Dead Sea Rift and the Gulf of California, have potential for altering sea level when they desiccate and then flood, but only slightly because their volumes of water are small in proportion to the volume of the ocean as a whole. Although irregular through geologic time, the occurrence of such events involving the origin of basins, their filling with water or drying up, is more likely during times of continental fragmentation and plate rearrangements.

As continental blocks moved about tectonically, ice centers waxed and waned irregularly, constituting one of the causes of cyclothems (Wanless and Shepard, 1936; Crowell, 1978; Heckel, 1986; Youle et al., 1994). For example, the Late Paleozoic Ice Age lasted for some 82 m.y. For more than 400 m.y. the united Gondwana supercontinent glided across the South Pole and high-latitude ice centers migrated with it during the late Paleozoic, as well as during the older Late Devonian–early Carboniferous and Ordovician-Silurian ice ages (Crowell 1983b; Caputo and Crowell, 1985). For earlier times, plate-tectonic reconstructions still lack enough veritability to be of great value in drawing inferences concerning the location of ancient glaciers and their debris, but progress is underway. The changing arrangements of continents and oceans is a prime control on climate throughout geologic time (Crowell and Frakes, 1970; Crawford and Daily, 1971; Frakes and Kemp, 1972; Frakes and Francis, 1988, 1990). This preview of methods used to document and to describe ancient glacial history and factors needing consideration is provided here as an entry to descriptions of the major ice ages in the following pages.

LATE PALEOZOIC ICE AGES

Glaciers reached sea level around the margins of Gondwanan continents during late Paleozoic time, waxing and waning for some 82 m.y. between ca. 256 Ma and 338 Ma. The glaciations largely died out about 5 m.y. before the end of the Permian period, although in Queensland alpine glaciers probably lingered. The end of the Permian is placed here at ca. 251 Ma (Claoué-Long, 1991; Veevers et al., 1994a; Ross et al., 1994; Yugan et al., 1997) in accord with recent U/Pb zircon dating of 251.4 ± 0.3 Ma from ashbeds on either side of the boundary in China (Bowring et al., 1998). Gradstein and Ogg (1996), however, place it at ca. 248 Ma and Roberts et al. (1996) at 245 Ma. Recent SHRIMP zircon dating in eastern Australia indicates that the base of the Kiaman reversed magnetic superchron is at ca. 318 Ma (Opdyke et al., 1999). This is above the Mississippian-Pennsylvania boundary, located at ca. 323 Ma by Harland et al., (1990). Accordingly, the top of the superchron is at ca. 263 Ma and suggests ca. 245 Ma as the ending date for the Permian period based on tentative correlations between Australia and the type Permian section in Russia (Roberts et al., 1996). This is 6 m.y. younger than the 251 Ma date used here following Harland et al. (1990) and Bowring et al. (1998).

The Late Paleozoic Ice Age began in mid-Viséan time at ca. 338 Ma and glaciers reached their widest spread between and ca. 320 Ma (approximately mid-Namurian) and ca. 256 Ma (Tatarian-Kazanian transition). Major repetitions or cycles in the glacial history are recorded by the stratal record and provide information on both climate and tectonic events, as discussed below. Before this major ice age began, there was a warmer interval of about 15 m.y. following upon the retreat at ca. 353 Ma of early Carboniferous ice in the early Viséan. In addition to the Gondwanan record, iciness occurred in Laurasian regions, but perhaps there were no large glaciers on land. The name "Late Paleozoic Ice

Age" is applied here rather than "Permo-Carboniferous" because iciness is documented from well back in the early Carboniferous (approximately mid-Mississippian = mid-Viséan) to late in the Permian and is not restricted to the time of transition between the late Carboniferous (= latest Pennsylvanian) and the Permian.

The sequences of waxings and wanings show that the late Paleozoic witnessed one of the most intense ice ages in Earth history but different in tectonic settings and mixes of causes from those of late Cenozoic and Late Proterozoic times. Because the late Paleozoic glaciations both began and ended during the Phanerozoic, geoscientists have an opportunity to study the comparatively complete record and learn why the glaciations began and why they ended. Such studies are vigorously underway and bear on the reasons for the marked end-Permian extinction and its biotectonic causes (Renne and Basu, 1991; Renne et al., 1992, 1995; Erwin, 1993, 1994; Wang et al., 1991, 1994; Knoll et al., 1986; Bowring and Erwin, 1998; Bowring et al., 1998). The Late Cenozoic Ice Age, in contrast, has not yet ended—we are now living well within it and perhaps near its culmination and can only speculate about its eventual ending.

Gondwanan glacial record

A glacial record occurs on all Gondwanan continents including portions that have been carried by tectonic movements from near-polar southern regions into the Northern Hemisphere after fragmentation of the supercontinent (Du Toit, 1921; Martin, 1961, 1981c; Wanless and Cannon, 1966; Crowell and Frakes, 1975; Crowell, 1983b, 1995; Caputo and Crowell, 1985; Chumakov, 1985; Veevers and Powell, 1987, 1994; Frakes et al., 1992; Eyles, 1993; Veevers, 1994a; Veevers et al., 1994a,b,c,d; González-Bonorino and Eyles, 1995; Veevers and Tewari, 1995). During maximum ice spread, large ice sheets lay upon united Gondwana centered near the join between southern Africa, Antarctica, and South America, and ice caps also thrived in then-attached Australia and India. The glacial episodes are documented by striated pavements someplace on Earth, but especially by widespread deposits of glacial debris, including faceted and striated megaclasts preserved in thin-bedded basinal sediments of continental, lacustrine, and marine facies (Figs. 27–29).

Africa. In southern Africa (i.e., South Africa, Namibia, Zimbabwe, Angola, and Mozambique) several advances and retreats of late Paleozoic ice sheets are documented. These shed glacial debris into the Karoo and Kalahari basins (Du Toit, 1921; Martin, 1961, 1981a,b; Frakes and Crowell, 1970; Crowell and Frakes, 1972; Rocha-Campos and De Oliveira, 1972; Rust, 1975; Talbot, 1981; Von Brunn and Stratten, 1981; Visser, 1981, 1982, 1987, 1990, 1993a,b, 1994a,b, 1997a,b; Visser and Loock, 1982; Von Brunn and Gravenor, 1983; Von Brunn 1987, 1994; Veevers, 1988; Von Brunn and Marshall, 1989; Eyles, 1993; Veevers et al., 1994c; Veevers and Tewari, 1995; Visser et al., 1997). Along now-southeastern Africa, glacial tongues entered from then-attached Antarctica and contributed sediment to the Karoo basin in what was probably part of a backarc basin associated with tec-

Figure 27. Striated side of *roche moutonnée* with Permian Dwyka tillite in depositional contact. Nooitgedacht Preserve, Kimberley District, South Africa. Point on 15-cm scale shows ice-flow direction. February 1967. Republished from Crowell and Frakes (1972, fig. 12).

Figure 28. Striated and faceted quartzite boulder, ~140 cm long, in Permian Dwyka tillite. Onrust Farm, east of Sand River and south of road between Calvert and Gluckstadt, ~25 km south-southwest of Vryheid, Natal, South Africa. February 1967. Republished from Crowell and Frakes (1972, fig. 9).

tonic convergence along the Panthalassic margin of Gondwana (Fig. 30). Lobes also entered southern Africa from highlands north of the Karoo basin and the Kalahari basin even farther north, and from highlands extending along the general trend of the African rift system. In tropical and equatorial Africa at present, patches of upper Paleozoic glacial strata are found in Gabon and Zaire (Cahen and Lepersonne, 1981a, b; Hambrey,

Figure 29. Striated veneer of Permian Dwyka tillite lying depositionally on locally polished, faceted, and striated Precambrian Ventersdorp Lava. Nooitgedacht Preserve, Kimberley District, South Africa. Opened knife has a total length of 14.5 cm with point of blade showing direction of ice travel. February 1967.

(McClure, 1978, 1980; McClure and Young, 1981; Al-Laboun, 1987; McGillivray and Husseini, 1992), at the border between Egypt and Sudan west of the Red Sea (Klitzsch, 1983), in Oman (Braakman et al., 1982; Levell et al., 1988; Robertson et al., 1990), in Yemen (Kruck and Thiele, 1983; El-Nakhal, 1984), and in Ethiopia (Dow et al., 1971; Hambrey, 1981a; Worku and Astin, 1992).

Facies studies in southern Africa reveal complex intertonguings where coarse tillites in terrestrial environments lie upon an irregular basement topography and finger into marine basinal deposits (Visser, 1997b). The sandstone, shale, and diamictite layers, along with other rock types, are assigned to the Dwyka Formation. Correlation from stratigraphic section to section is aided in the semiarid region because units can be followed from place to place in the field, supplemented by information from boreholes and other subsurface data. Facies, such as those characterizing the types of diamictite, are first traced out before arriving at interpretations of depositional processes and environment. These interpretations are largely based upon a comparison with those of the late Cenozoic where processes and products associated with sedimentation beneath and around ice sheets are better known. For example, thinly bedded and sheared diamictite may be the result of deposition associated with subglacial ice streams where the temperature at the time of deformation and deposition was close to the ice melting point (e.g., Anandakrishman et al., 1998; Bentley, 1998; Bell et al., 1998). No diamictites in the Dwyka have been recognized as the result of bolide impacts (Reimold et al., 1997).

In southern Africa, and extending northward, three major cycles are recognized in the facies and mark episodes in the

1981b), Central African Republic (Visser, 1997b), Tanzania (Wopfner and Kreuser, 1986; Kreuser et al., 1990; Wopfner and Casshyap, 1997), Zambesi (Bond, 1981a), Zimbabwe (Bond, 1981b), and in Madagascar (Frakes and Crowell, 1970; Wopfner and Casshyap, 1997). These remnants are largely continental and lacustrine in facies except perhaps for those in the Congo basin on the west where marine incursions are suspected.

In the Northern Hemisphere of Africa at present, Gondwanan glacial beds are found in central and southern Arabia

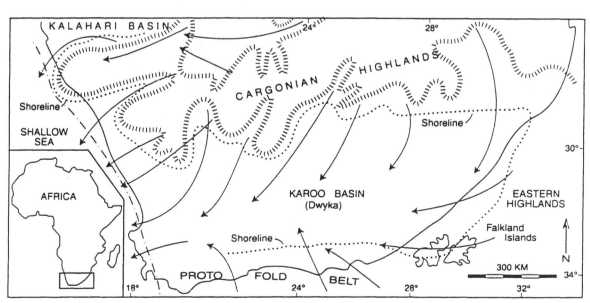

Figure 30. Southern Africa and then-attached regions during Permian maximum glaciation between ca. 277 Ma and 290 Ma (Early Permian). Long arrows depict general pattern of ice flow. See text for discussion of position of Falkland Islands, which here are displaced and rotated far to the east of their present position that resulted from the opening of the Southern Atlantic Ocean. Dash and dot line shows western maximum limit of ice margin at about the beginning of the Permian (Visser, 1997a, fig. 1). Simplified and modified from Veevers et al. (1994b, Fig. 11).

deglaciation history of Gondwanan ice sheets (Visser, 1997b). The oldest cycle took place in late Westphalian–Stephanian time (ca. 295 Ma to ca. 305 Ma), an intermediate cycle during very late Carboniferous–early Sakmarian time (ca. 278 Ma to ca. 292 Ma), and the youngest during late Sakmarian–Artinskian time (ca. 272 Ma to ca. 275 Ma). Ice sheets over the region waxed and waned irregularly during these late Carboniferous and Early Permian episodes, but the time gaps between sequences and the timings of the intervals are not well known. Maps depicting the interpreted paleogeography at these times show marked variations in topography, location of ice centers, and size and shape of depositional basins offshore from ice tongues extending to the sea (Visser, 1997b). Not only is high latitude positioning of the supercontinent deemed responsible for the glaciation, but so are the levels of the sea, local topography, regional tectonics that bring about mountains and basins, and availability of open water for evaporation.

Sediments are viewed as reaching from the Karoo basin of southern Africa into the Sauce Grande basin of then-attached Argentina, and from the Kalahari basin into the Chaco-Paraná and Paraná basins of Brazil, Uruguay, and adjacent Argentina (Visser, 1997b; López-Gamundi, 1987, 1989, 1991, 1997). In central Africa, glacial facies, correlated with the Dwyka Formation to the south, display subtle coarse-to-fine tongues associated with rising rift shoulders, suggesting that the mid-African rift system was nascent. The system of Permian rifts, with glaciers at times upon their mountainous shoulders, can be followed across united Gondwana well into Asia (Wopfner and Casshyap, 1997).

South America. Glaciers thrived in southern South America during late Paleozoic time (Fig. 31). The Gondwanan glacial beds occur especially in the Paraná basin (Rocha-Campos, 1967; Frakes and Crowell, 1969; França and Potter, 1991; Eyles et al., 1993; França, 1994; López-Gamundi, 1997; Ziegler et al., 1997; López-Gamundi and Rossello, 1998), in basins in western Argentina (López-Gamundi, 1987, 1989, 1991, 1997; Eyles and Eyles, 1993; Eyles, 1993; López-Gamundi et al., 1994; Dos Santos et al., 1996), in the Sierra de la Ventana of southeastern Argentina (Coates, 1969; Amos and López-Gamundi, 1981), in the Falkland Islands (Frakes and Crowell, 1967; Greenway, 1972; Marshall, 1994) and in Bolivia (Helwig, 1972; Fernandez Garrasino, 1981). During times of maximum glaciation, sediment arrived in eastern South America from sources in then-attached southern Africa.

An early glaciation of late Namurian–early Westphalian age (ca. 320 Ma to ca. 308 Ma) is documented primarily in western South American basins near proto-Andean mountains. These basins were probably parts of a long chain of basins adjoining a convergent plate boundary. Downslope sliding prevailed along this topographically irregular tectonic belt and carried glacial debris along with much material from other sources into nearby basins, but the interpretation that some of the debris is glacial has been questioned (Cortelezzi and Solís, 1982). Rather than huge continental ice sheets, paleogeographic reconstructions suggest irregular basins with uplands between them upon which local

glaciers and ice caps occurred. A later episode of glaciation of late Stephanian to early Sakmarian age (ca. 278 Ma to ca. 292 Ma) is recognized in several basins, especially those on the southeast and correlated with those in southern Africa. For this later glaciation as well, the topography is reconstructed as irregular, with moderately high cratonic regions separated from deep basins.

The glacigenic deposits of the Paraná basin, along with the Chaco-Paraná basin to the west, are well documented in view of good intermittent exposures supplemented by much well data (Dos Santos et al., 1996). Paleo–ice flow directions determined from striations on basement exposures and from paleocurrent structures and facies changes trend from the southeast toward the northwest and not from the interbasinal highs. Isopachs show sediment bodies at the Paraná basin margins extending westward into these basins as well as into the Sauce Grande basin at the south. These lobes reaching basinward and westward include, from north to south: Kaokoveld, Paraná, and Uruguay, and are viewed as extensions of lobes in southern Africa (Dos Santos et al., 1996). A glacial record that probably correlates with these deposits is reported from Angola (Rocha-Campos and De Oliveira, 1972).

The Falkland Islands contain a glacial record preserved in the Lafonian Diamictite and associated facies (Frakes and Crowell, 1967; Greenway, 1972; Marshall, 1994). Stones within massive diamictites mainly consist of rock types present in the underlying Devonian sequences but also contain huge boulders of basement rock up to 7 m in size, similar to rocks exposed on Cape Meredith, the southern tip of west Falkland. An ice cap is inferred to have lain to the present west, with ground moraine and eskers at its margin. This gave way eastward to a floating ice shelf with local sediment fan lobes beneath it and extending into deeper water above a slope down which debris sliding took place. Farther to the east, open water is inferred from dropstones within thin-bedded turbidites.

Plate tectonic reconstructions, based largely on paleomagnetic data but also evaluating stratal and rock correlations, place the Falkland Islands originally to the southeast of southern Africa (Mitchell et al., 1986; Thistlewood et al., 1997). The island block is viewed as having been moved far to the west from its original African site within a complex system of transform faults and rotated 120°. These displacements occurred during the breakup and dispersal of Gondwanan segments in Mesozoic time when the South Atlantic Ocean opened—a reconstruction originally advanced by Adie (1952). Cape Meredith rocks are deemed to be similar to basement complexes in Natal. The West Falkland ice cap or sheet is placed as part of the glacier system to the north of the African Cape foldbelt.

Australia. The late Paleozoic glacial record is also widespread in Australia (Figs 32–34) (Crowell and Frakes, 1971a,b; Dickins, 1976, 1985, 1993, 1997; Quilty, 1984; Veevers and Powell, 1987; McPhie, 1987; Brakel and Totterdell, 1992; Veevers et al., 1994a,c,d; Veevers and Tewari, 1995; Lindsay, 1997). Many glacigene facies are recognized, ranging from subglacial deposits outward from land sources to deep marine sequences with dropstones. The main ice centers contributing glacial debris

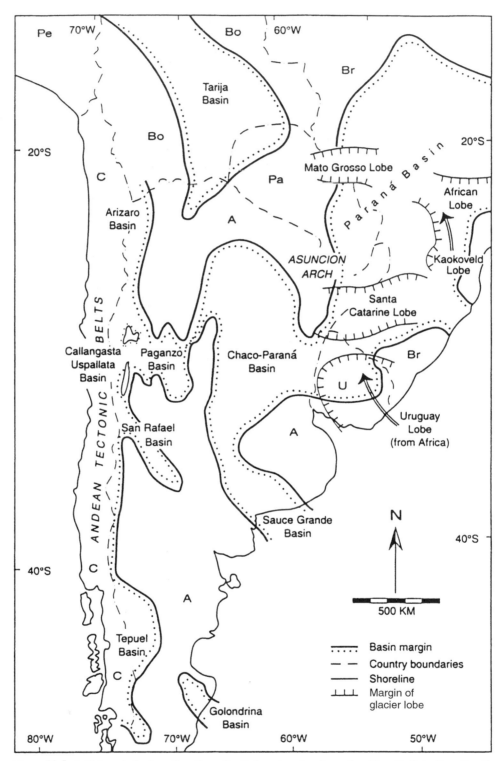

Figure 31. Late Paleozoic basins of southern South America. A = Argentina; Bo = Bolivia; Br = Brazil; C = Chile; Pa = Paraguay; Pe = Peru; and U = Uruguay. Simplified from França and Potter, 1991, fig. 7; López-Gamundi, 1997, fig. 9.2).

to Australia lay in highlands bordering Antarctica and flourished long before the separation of Australia from Antarctica. The glacial record is also preserved around the margins of the Western Australian Yilgarn and Pilbara basement blocks, and in South Australia, Victoria, and Tasmania. Ice caps are documented as well in central Australia, and as far north as the Kimberley region, a region that provided ice rafted debris to the Bonaparte basin. The Central Australian Uplands are depicted by Veevers (1994b) as shedding ice outwardly, including southward into a lowland separating ice and fluvial flow from Antarctica.

In Australia, glaciation began in late Carboniferous time (approximately late Namurian = ca. 310 Ma) when the eastern margin of Australia was part of the convergent plate margin of Gondwana and consisted of mountain ranges and nearby deep basins, including those of the New England foldbelt (Brakel and Totterdell, 1992; Veevers et al., 1994a; Lindsay, 1997). The Queensland ranges and associated coastal basins include an arc system bordering the Panthalassic Ocean. Glaciers shed debris into basins as far north as present 20° lat, and in Australia as far south as

Tasmania. In the northern Bowen basin of Queensland, large dropstones in sequences as young as ca. 254 Ma (Kazanian stage) are viewed as perhaps derived from alpine glaciers coming down to the sea from ranges not far to the west, or they may have drifted north in floes carried by oceanic currents parallel to the coast and from sources well to the south, including then-attached Antarctica. Following upon the initiation of glaciation among mountains of eastern Australia, ice centers expanded and culminated in central and Western Australia in Early Permian time (approximately Sakmarian = ca. 278 Ma). Local alpine glaciers within the Queensland ranges probably lasted to the end of the Permian.

Dating of the Australian glaciations depends upon fossil leaves of Gondwanan floras, spores, pollen, marine invertebrate faunas, and stratigraphic position. Isotopic dating is available mainly in the complex Queensland region from volcanic rocks, including tuffs (Veevers et al., 1994a).

Antarctica. Glacial strata of Carboniferous and Permian age are found in several regions of Antarctica (Fig. 35) (Lindsay, 1970; Frakes et al., 1971; Elliot, 1975; Ojakangas and Matsch,

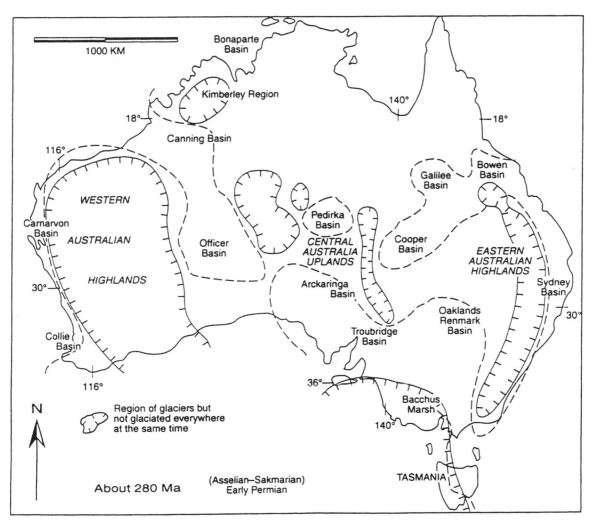

Figure 32. Highlands and basins of Australia during the Early Permian at ca. 280 Ma showing regions of glaciers upon highlands, although these were not glaciated everywhere at the same time.

Figure 33. Hollow in Permian glacially striated and sculptured pavement containing preserved tillite. Pyrete Creek, near Coimadai, Bacchus Marsh area, Victoria, Australia. April 1968. Republished from Crowell and Frakes (1971a, fig. 6).

Figure 34. Permian dropstone boulder upon crowded brachiopod assemblage. Maria Island, Tasmania. May 1968.

1981; Coates, 1985; Miller and Waugh, 1991; Miller, 1989; Barrett, 1991; Matsch and Ojakangas, 1992; Collinson et al., 1994; Veevers and Collinson, 1994; Isbell et al., 1997). They crop out along the chain of the Transantarctic Mountains and in the Sentinel Range (Ellsworth Mountains). The glacial beds, assigned to several formations, at places lie upon glacially striated pavements and include massive diamictites containing striated stones (Fig. 35) and thin-bedded units with dropstones. They are viewed as extending from subglacial lodgment deposits to marine beds laid down below a floating ice shelf. Several advances and retreats

are recognized that correlate satisfactorily with Australian sequences from Early Permian (Sakmarian? = ca. 278 Ma) back into the early Carboniferous (Late Mississippian, perhaps within the Namurian = ca. 320 Ma). Mountain ranges in Antarctica during mid-Carboniferous time rose and partly controlled fluviatile drainage into then-attached Australia (Veevers, 1994b; Veevers and Tewari, 1995).

The belt of strata including those of glacial derivation is interpreted as lying upon the margin of the wide East Antarctic craton (Collinson et al., 1994; Veevers and Tewari, 1995; Isbell et al., 1997). The strata are viewed as deposited along the Panthalassic margin of Gondwana within a broad belt, including forcland basins, that was later the site of the Gondwanide fold-thrust belt formed largely during Mesozoic time. The mountain range was still later uplifted during Cretaceous and early Tertiary time resulting in the exposure of the beds by erosion today (Stern and Ten Brink, 1989; Stump and Fitzgerald, 1992). The thickest section of glacigene beds occurs within the Ellsworth Mountains, which at the time of deposition probably lay along the trend of the present Transantarctic Mountains north of the Pensacola Mountains (Fig. 36) (Grunow et al., 1991, 1996). This interpretation calls upon postdepositional tectonic rotation of the Ellsworth block of approximately 90° and its lateral translation to bring it into rough alignment with the Panthalassic trend. Mapping, dating, and correlation along the belt so far do not allow a decision whether ice covered the whole upland region bordering the Panthalassic margin at any one time, now part of East Antarctica, or whether ice centers migrated along this >4,000-km-long uplifted belt. During the late Paleozoic, Antarctica was joined to southern Africa at one end and to Australia at the other.

Asia. In India tillitic debris, assigned to the Talchir Formation, overlies glacially striated pavements at the margins of several elongate basins or rifts within the Precambrian craton. These basins include the Damodar, Son, Chotanagpur, Mahanadi, and Pranhita-Godavari (Fig. 37) (Frakes et al., 1975, 1976; Schwarzbach, 1976; Casshyap and Kumar, 1987; Casshyap and Srivastava, 1987; Veevers and Tewari, 1995; Dickins and Shah, 1987; Dickins, 1997; Wopfner and Casshyap, 1997). Subsidence of the basins, accompanied by deposition of the Talchir Formation, took place near the end of the Carboniferous period (ca. 290 Ma) and is apparently associated with a developing rift system. It is not yet known how widespread the glaciers were at that time, or whether they were limited to rift shoulders or margins. The tillitic debris was followed by deposition of widespread strata of nonglacial facies assigned by Veevers (1994b) and Veevers and Tewari (1995) to the Gondwana master basin of peninsular India as the rifts filled and their margins were overlapped. Reconstructions by Veevers (1994b, fig. 2A) show alignment between the Lambert Graben in Antarctica and the Mahanadi rift in India, an alignment that controlled drainage northward at the beginning of the Permian.

Glacial deposits along the northern border of Gondwana are now incorporated into the tectonically complex Himalayan belt (Fig. 38) (Singh, 1987; Şengör, 1987; Şengör et al., 1988; Sun Dong-li, 1993; Veevers and Tewari, 1995). This belt extends

Figure 35. Permian-Carboniferous striated boulder in Whiteout Conglomerate, near Mt. Earp, Sentinel Range, Ellsworth Mountains, Antarctica. Scale: Centimeters below, inches above. December 1966.

northwest into Pakistan and Tibet and southeast into Burma, Malay, and Thailand (Frakes et al., 1975; Altermann, 1986; Stauffer and Lee, 1987; Metcalfe, 1988). At the time of the glaciations, before the northward movement of the Indian subcontinent that collided with mainland Asia and other tectonic blocks to construct a complex tectonic belt now largely within the Himalayan belt and Tibetan Plateau, the glacial centers bordered the paleo-Tethyan seas. These seas provided evaporative moisture to precipitate as snow along the fringe of Gondwana.

The Laurasian icy record

In the Northern Hemisphere during late Paleozoic time, cold climate in Siberia and arctic Canada is documented by beds containing small dropstones, but the debris was probably carried by shore or river ice, not necessarily by true glacial icebergs (Frakes et al., 1975; Stanley, 1988; Beauchamp et al., 1989; Beauchamp, 1994). In the Verkhoyansk region of Siberia marine facies with small dropstones of Permian age are reported (Epshteyn, 1978). Both the Siberian and Canadian (Sverdrup basin) localities include strata with cold-water fossils and are overlain at places by coal beds of uppermost Permian age (Tatarian stage, ca. 251–265 Ma).

Discussion: Late Paleozoic ice ages

The late Paleozoic glaciations occurred between ca. 338 Ma and ca. 256 Ma as the Gondwanan supercontinent drifted across the South Pole and began to merge with Laurentia to form Pangea (Crowell and Frakes, 1970; Crowell, 1983b; Caputo and Crowell, 1985; Scotese, 1990; Klein, 1994; Veevers, 1994a). Its movements brought about a shifting of climate belts and changes in the flow patterns of air and ocean currents (Parish, 1993;

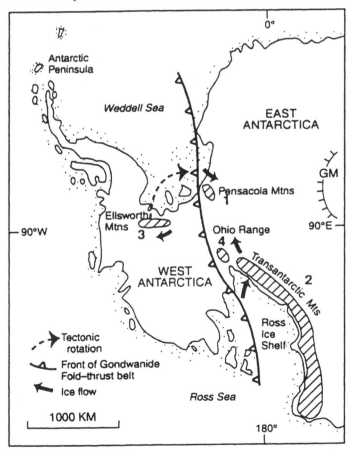

Figure 36. Permian-Carboniferous glaciated regions in Antarctica. (1) Gale Mudstone; (2) Pagoda Formation, Metschel tillite, Darwin tillite; (3) Whiteout Conglomerate, including tillite; (4) Buckeye tillite. GM = Gamburtsev Mountains. Simplified from Grunov et al. (1991, fig. 5a); Miller and Waugh (1991, fig.1); Eyles (1993, figs. 16–18).

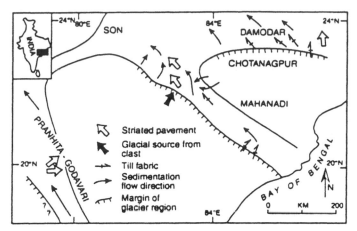

Figure 37. Permian glaciated regions, east-central India. After Veevers and Tewari (1995, fig. 20).

Figure 38. Permian metadiamictite, interpreted as a metatillite. Sethi-Khola tillite. Darjeeling region, India. Coin is about 1.5 cm in diameter. January 1977.

Klein, 1994; Veevers, 1994a). The supercontinent stood high relative to sea level and probably lay upon a thick, deep foundation consisting of a large thermal bulge (Veevers and Powell, 1987). Topography upon Gondwana was irregular and related to convergent and transform plate-tectonic margins around it and to nascent rift systems cross-cutting it from India into southern Africa (Daly et al., 1991; Veevers, 1994a; Wopfner and Casshyap, 1997). When viewed on Gondwanan reconstructions, ice extended away from the pole in the Southern Hemisphere as far as 40° lat in the Early Permian (Nie et al., 1990; Ziegler, 1990; Powell and Li, 1994; Veevers et al., 1994c). Ice reached northward to about the same latitude as Pleistocene glaciers reached southward in North America.

The dating of the late Paleozoic glaciations depends primarily on plants, spores, and pollen from terrestrial interbeds, including coal beds within associated and overlying strata. Locally, these beds at the margins of ice centers grade into marine facies with fossils. The age assignments are therefore dependent on adequate lateral facies correlations and understanding of faunal and floral ranges. Gondwanan plants, such as *Glossopteris,* as well as spores and pollen occur widely (Truswell, 1980; White, 1990). Marine beds carrying fossils such as *Eurydesma* and *Levipustula* associated with the glacial horizons are also found, especially in southwestern Africa, South America, Australia, and Antarctica, and have strengthened correlations (Dickins, 1985, 1993; Visser, 1990). Many paleontologists have now contributed to correlations between Laurasian and Gondwanan sequences although there is still controversy over calibration with time scales (Ross and Ross, 1985, 1988; McKerrow and Scotese, 1990; Beauchamp, 1994; Klein and Beauchamp, 1994; Veevers, 1994a; Veevers et al., 1994a,b; Roberts et al., 1996; Gradstein and Ogg, 1996). There are only a few localities where interbedded volcanic layers allow isotopic dating, such as in northeastern Australia (Veevers et al., 1994a; Roberts et al., 1996; Opdyke et al., 1999).

The last evidence of the Late Paleozoic Ice Age, consisting of ice-rafted debris, is recorded in strata of Kazanian age (ca. 256 Ma) in eastern Queensland, Australia. Here ice-rafted boulders as much as 2 m across are found today as far north as 20° S lat within deposits laid down in basins along the Gondwanan continental margin with Panthalassa where volcanic arcs occurred on the west (Brakel and Totterdell, 1992). Large dropstones, consisting of basement types, may have been derived from glaciers incising the craton on the west, or carried by ocean currents from glacial sources far to the south and even as far as from then-attached Antarctica ~2,300 km away (Crowell and Frakes, 1975; Crowell 1983b; Veevers et al., 1994a). In the Canadian arctic fossils document a cooling trend in approaching the end of the Permian (Stanley, 1988; Beauchamp, 1994).

Under the premise that ice accumulated in high latitudes where continental margins or mountain belts acquired snow that lasted from season to season, the distribution of ice centers fits the concept that glaciers waxed and waned in some kind of sequence as the united supercontinent moved across the South Pole and as the tectonic positions of uplands and basins changed (Figs. 39–41) (Crowell 1983b; Caputo and Crowell, 1985; Eyles, 1993). Ice centers began to grow in now-western South America in late Viséan time (ca. 338 Ma), and then moved now-eastward to develop their maximum spreads in regions adjoining southern Africa during Westphalian to Artinskian time (from ca. 310 Ma to ca. 272 Ma). The ice age diminished through Kungurian time as the ice centers occupied Australia and then-adjoining India, and died out in the early Kazanian at ca. 256 Ma in now-northeastern Australia ca. 5 m.y. before the end of the Permian period. Only local small alpine glaciers in the coastal mountains of Australia are presumed to have remained.

The tracking of ice masses across the Gondwana supercontinent, as shown by dating of the preserved glacial record, documents the movement of the supercontinent across high latitudes with a nearby south pole (Crowell, 1983b; Caputo and Crowell, 1985; Veevers et al., 1994c). Although the facies record and its

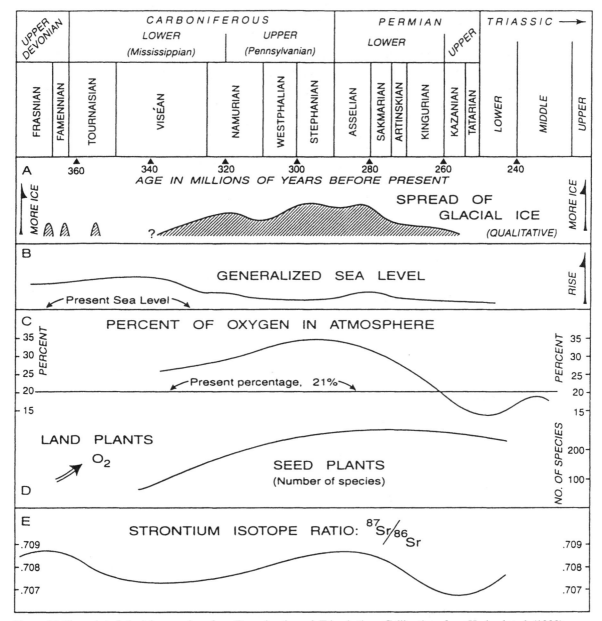

Figure 39. Time plot of glacial expansions from Devonian through Triassic time. Calibrations from Harland et al. (1990), generalized sea-level curve from Vail et al. (1977), percent of oxygen in atmosphere from Berner (1990), number of species of land plants from Niklas et al. (1985), strontium isotopic ratios simplified from Burke et al. (1982) and Holser et al. (1988). Modified from Crowell (1995, fig. 2).

interpretation are independent of paleomagnetic measurements, the paleomagnetic data confirm and reenforce the wander path of the supercontinent (e.g., Bachtadse and Briden, 1990). The drift of Gondwana is therefore supported independently by *both* paleomagnetic interpretations *and* the reconstruction of glacial facies and their ages under the premise that sea-level glaciers occur primarily in high latitudes. This underpins the hypothesis that ice ages as well as other major climate changes on Earth are, in part if not largely, controlled by global movements of continental cratons and associated changes in land topography and ocean bathymetry. Inasmuch as this concept is satisfactorily established

for the late Paleozoic, the hypothesis is logically extended back in time to explain some earlier climate changes.

The spread and timing of ice accumulations show that the whole of Gondwana was never glaciated at the same time (Figs. 40, 41) (Roberts, 1976). Time-space diagrams, carefully constructed by Veevers et al. (1994c), show a widespread lacuna (an unconformity of both erosional and nondepositional origin) in the Carboniferous. In fact, an explanation for this lacuna remains as a major paleotectonic problem although it is likely that the lacuna results from high-standing Gondwana sitting upon a thick lithospheric or upper-mantle bulge. In South America and northeastern Australia, however, glacial deposits were laid down during this interval. These are assigned to a "Glacial I" episode ranging between ca. 290 Ma and ca. 338 Ma. In eastern Australia, "Glacial I" merges with a "Glacial II" interval, which extended widely around Gondwana in the Early Permian between ca. 277 Ma and ca. 330 Ma. There is little relation between the extent of the preserved glacial record and the interpreted extent and thickness of glaciers on land (González-Bonorino and Eyles, 1995). The record preserved for us to study is likely to have been laid down in a basin offshore. There is no evidence to suggest that sea level was lowered more than ~350 m or approximately the same as the maximum worldwide lowering documented for the Late Cenozoic Ice Age (Crowley and Baum, 1991a, 1992; Crowley et al., 1991). If the whole of Gondwana had been under ice at any one time, the lowering would presumably have been much greater than this amount. As the supercontinent drifted and the mass of glacial ice increased, oxygen increased concurrently to percentages significantly greater than those of today (Fig. 39) (Berner, 1990), and carbon dioxide decreased. Land plants flourished and added oxygen to the air.

The Late Paleozoic Ice Age ended, presumably as the result of several major geotectonic and associated events. The supercontinent of Pangea (the supercontinent that resulted from the merging of Gondwana and Laurentia) began to subside and fragment and sea level rose. Plate-tectonic activity increased and continental blocks are viewed as having moved about more vigorously. The paleo-Pacific (Panthalassan) coast of Gondwana rotated and moved away from the South Pole (Crowell, 1978, 1995). Ocean and air currents were now able to circulate more freely and zonally (latitudinally) about the Earth's spin axis in polar regions, and were redirected longitudinally elsewhere by land and sea rearrangements. Biogeochemical modifications, such as changes in the carbon flux to shelf waters and to the air increased. At the transition of the Permian into the Triassic, marked shifts occurred in the carbon, sulfur, strontium and other geochemical ratios that are interpreted as related to these changes (Burke et al., 1982; Holser et al., 1988; Gruszczyński et al., 1989; Hudson, 1989; Raymo 1991; Magaritz et al., 1992; Erwin, 1993, 1994; Wang et al., 1994; Knoll et al., 1996; Bowring and Erwin, 1998; Bowring et al., 1998). Widespread

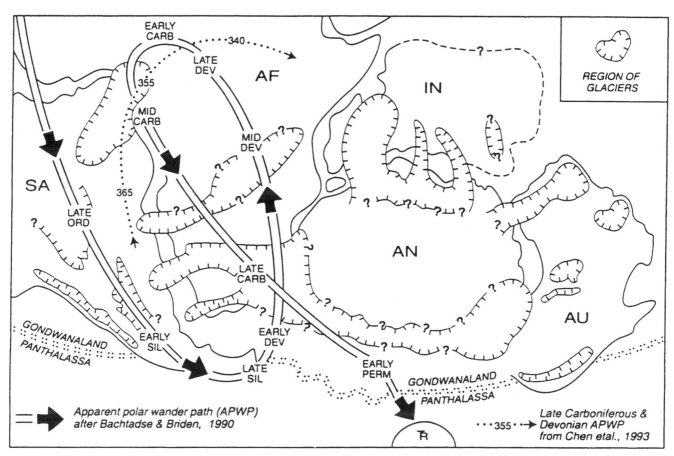

Figure 40. Gondwanan glaciated regions from Ordovician to Permian time with apparent polar wander paths from paleomagnetic data after Bachtadse and Briden (1990) and Chen et al. (1993). Numbers along Chen et al. apparent polar wander path in millions of years. AF = Africa; AN = Antarctica; AU = Australia; IN = India; SA = South America; CARB = Carboniferous; DEV = Devonian; ORD = Ordovician; PERM = Permian; SIL = Silurian. Simplified and with additions from Crowell (1995, fig. 4).

Figure 41. Southern Hemisphere glaciated regions during Early Permian (Artinskian at ca. 260–269 Ma). AF = Africa; AN = Antarctica; AU = Australia; SA = South America. Projection and Gondwanan assembly modified from Irving (1983, fig. 3). From Crowell (1995, fig. 3).

Figure 42. Late Devonian–Early Carboniferous basins with a glacial record. AF = Africa; SA = South America; NA = North America. A = Acre basin; B = Solimões basin; C = Amazonas basin; D = Parnaíba basin; E = Paraná basin; F = Proto-Andean basins; G = African basins. Continental reconstruction from Kent and Van der Voo (1990, fig. 6, Famennian).

anoxia in the oceans is viewed as contributing to the extinction (Wignall and Twitchett, 1996; Isozaki, 1997). Strontium isotopic ratios ($^{87}Sr/^{86}Sr$) show an increase as the glaciers culminated and a decrease as they waned. Lower strontium ratios are interpreted as the result of two processes: (1) Increased tectonic activity along mid-ocean ridges brought more primitive ^{86}Sr from deep within the mantle. Lower values suggest increased worldwide tectonism, active volcanism, and hydrothermal activity. (2) Widely exposed ancient continental crust at high elevation contributed to higher Sr isotopic ratios. Inasmuch as continents on average were high and were capped with glaciers, erosion rates on land were also correspondingly high. Rivers carrying strontium to the sea were rich in the more radiogenic strontium isotope. As continents were lowered, sea level rose and flooded continental shelves decreasing the amount of plant and soil CO_2. These and other factors contributed to increased CO_2 in the atmosphere. The interplay of these many tectonic, continental, positional, biogeochemical, and oceanographic factors are viewed as prime forces in accounting for the end of the ice age, an ice age that fades away as the Mesozoic era approaches.

The record shows that widespread continental glaciation ended some 5 m.y. before the marked extinction at the very end of the Permian and is associated with strong geochemical changes (Magaritz et al., 1992; Erwin, 1993, 1994; Wang et al., 1994; Retallack, 1995; Knoll et al., 1996b; Wignall and Twitchett, 1996; Isozaki, 1997; Bowring and Erwin, 1998; Bowring et al., 1998). New dating of zircons in tuffs interlayered with fossil-bearing beds bracketing the extinction shows that it took place over only a few hundred thousand years (Bowring and Erwin, 1998). During the 5 m.y. interval leading up to the extinction, there was voluminous volcanism including the outpouring of the Siberian traps (Renne and Basu, 1991) and from volcanic arcs along the Gondwanan-Panthalassic join in now-northeastern Australia (Veevers and Powell, 1994). Increased volcanism is viewed as significant by contributing much carbon dioxide to bring about rapid atmospheric and oceanic warming (Renne et al., 1992, 1995; Campbell et al., 1992). Because the warming was irregular, perhaps a cooling interval also led to a final glacial pulse (Stanley and Yang, 1994). Rapid, irregular warming and cooling of the atmosphere and ocean surface may have induced shifts in the oceanic conveyor-belt system and caused deep-water overturns. No bolide impact has been reported although Bowring et al. (1998) raise the possibility that an icy and carbon-rich bolide may have impacted Earth and triggered the extinction, but there is as yet no compelling evidence for such an event. Searches in Antarctica and Australia for evidence of an impact have been unrewarding (Retallack et al., 1998). The extinction itself, still under study by many geoscientists, may be associated with the overturn of anoxic deep water that flooded the shallow continental shelves as sea level rose. This is suggested by a marked addition of light carbon at the time of the mass extinction (Bowring and Erwin, 1998; Bowring et al., 1998). Even though the relative interplay of these many associated events near the end of the Permian are not yet satisfactorily understood, the root cause of both the ending of the

Figure 43. Glacial striations (oriented N 15° E) upon sandstone of the Devonian (Famennian stage) Cabeças Formation, Parnaíba basin, Brazil. About 1 km S of road from Canto do Burutí to Bom Jesus. September 1982. Republished from Caputo and Crowell (1985, fig. 8).

ice age and the later extinction probably lies in a complex combination of geotectonic, volcanic, and biogeochemical processes.

LATE DEVONIAN–EARLY CARBONIFEROUS ICE AGES

Over an interval of about 10 m.y. at the end of the Devonian and into earliest Carboniferous time, between ca. 353 Ma and ca. 363 Ma, two short episodes of confirmed glaciation, and perhaps a third, took place in the area now near the Amazon River in northern Brazil and then adjoining regions (Fig. 42) (Caputo, 1984, 1985, 1994, 1995; Crowell, 1983b; Caputo and Crowell, 1985). Glaciation documented in these regions is intriguing because the area now is equatorial in position and is covered in part by tropical rainforests. These circumstances emphasize either the mobility of crustal units or extreme changes in climate since such ancient times of iciness. The Late Devonian is a time of marked physical and biological changes, including the Frasnian-Famennian extinction (McGhee, 1996). It is also viewed as a cool

time in the midst of a long interval of warmth (Frakes et al., 1992). About 15 m.y. lapsed between the ending of the early Carboniferous icy episode and the beginning of the Late Paleozoic Ice Age.

Although diamictites observed in cores from the Amazonas basin were first interpreted as glacial by Moura in 1938, and several workers later added observations to document the likelihood of glaciation, it was not until 1971 that Caputo and Vasconcelos published data on striated clasts within the diamictites along with petrologic and field details to confirm a glacial origin. In the Parnaíba basin of northeastern Brazil, for example, striated pavements and striated megaclasts are exposed with associated tillites (Figs. 43, 44). These substantiate a glacial origin for diamictites that are widespread along basin margins and recovered from within the basin in cores from wells drilled for oil. In addition to the Parnaíba basin (Cabeças Formation), glacial strata are now known from two other basins along the Amazon trend, a trend that includes the Acre, Solimões (Jaraqui Formation), and Amazonas basins (Curiri and Lower Oriximiná Formations) (Caputo, 1985). The facies are dated on the basis of miospore assemblages (LN palinostratigraphic zone) that are deemed equivalent to the middle *Siphonodella praesulcata* zone at the top of the Famennian stage (Loboziak et al., 1992). This episode is correlated with diamictites of the Cumaná Formation within proto-Andean basins of Peru and Bolivia (Copacabana Peninsula and Lake Titicaca regions) (Vavrdová et al., 1991; Caputo, 1994, 1995; Díaz-Martínez and Isaacson, 1995). These are correlated with localities in Libya (Ashkidah and Tahara Formations, containing diamictites, of the Murzuq basin), in Algeria (Djebel Illèrene Formation of the Illizi basin), and in Ohio (Bedford Shale and Berea Sandstone) (Vavrdová et al., 1991).

Stratigraphic position shows that there were two distinct short glacial episodes within the Famennian, and perhaps an earlier third near the end of the Frasnian. Each episode lasted ca. 1 m.y. and was separated by somewhat longer intervals when ice did not reach outward from their sources on land to leave a stratal record. Ice may well have remained inland at places but if so, such a direct record has been lost through erosion. The slightly earlier and perhaps third glacial episode may be documented by diamictites in the Pimenteira Formation in the Parnaíba basin, but these strata have not yet been adequately dated.

The youngest glaciation or glacial advance is of mid-late Tournaisian age of the Carboniferous (between ca. 353 Ma and ca. 357 Ma) and is documented by tillites unconformably overlying Famennian glacial beds in the Solimões basin (Jaraqui Formation) and by diamictites in the Parnaíba basin (Poti Formation) (Loboziak et al., 1992). Miospore assemblages are correlated with the *Siphonodella crenulata* zone of the upper Tournaisian Stage. Tillitic beds of this age are also reported from Africa as far east as Libya where they can be fitted into standard sequences more satisfactorily (Caputo, 1995). The tentative dates of stages given here are adjusted between those listed in Palmer (1983) and in Harland, et al. (1990) and are subject to revision as new data become available.

Extinction events and paleogeographic arrangements during

Figure 44. Glacially faceted, striated, and polished cobble from Cabeças Formation (Upper Devonian, Famennian stage), western Parnaíba basin from east cutbank, Rio Tocantins, Brazil, ~20 km north of Pedro Afonso. Coin is 2.5 cm in diameter. September 1982. Republished from Caputo and Crowell (1985, fig. 10).

the Late Devonian, before and after glacial episodes, are receiving special attention (Heckel and Witzke, 1979; Wilde and Berry, 1984, 1988; House, 1985b, 1989; Johnson et al., 1985; McGhee, 1996). Although there were extinction events in Famennian time and they were strongly marked at its end, the Hangberg event immediately below the Devonian-Carboniferous boundary is also significant. This extinction was followed by the slow establishment and radiation of faunas with Carboniferous characteristics. The extinction at the end of the Famennian correlates well with the sea-level drop and glaciation in western Gondwana (Copper, 1977, 1986). This extinction probably took place when regression and cold-water circulation may have started due to ice-sheet buildup close to the end of Famennian time.

Although the major causes of the Late Devonian–early Carboniferous glaciations are deemed to lie in the polar positioning of the Brazilian part of Gondwana at that time, other events probably played a significant climate role. One or two bolide impacts may have occurred during the interval of glacial advances, although the dating is insecure (McLaren, 1970, 1983; Playford et al., 1984; Hurley and Van der Voo, 1990; McLaren and Goodfellow, 1990; Claeys et al., 1992; Wang, 1992; McGhee, 1996). In China stratal evidence supports an impact, but stratigraphic position suggests that the impact happened after the end-Frasnian extinction and before the end-Famennian extinction; whereas in Australia and Europe, an impact may have corresponded with the Frasnian-Famennian event. Chemical anomalies, including subtle iridium spikes, have been identified in Western Australia and south China but these anomalies may be due to biological processes rather than impacts (Hurley and Van der Voo, 1990). Microtektites (spherules of melt-glass thrown into the atmosphere from bolide impacts) are reported from China, Belgium,

and northwestern Canada (Boundy-Sanders, 1992; Claeys et al., 1992). As yet, time correlations do not establish how many bolide impacts occurred, if there were truly more than one. Moreover, the impact sites have not yet been identified, although this is unlikely because the sites may have been subducted long ago (e.g., McGhee, 1996).

In addition to the polar positioning at high latitudes of the affected parts of Gondwana, reductions in greenhouse gases are also viewed as important in causing glaciations in the latest Devonian and earliest Carboniferous (Caputo, 1995). Paleomagnetic studies show apparent polar wander paths as crossing over the region of join between Brazil and western Africa (e.g., Bachtadse and Briden, 1990) and with high drift velocities of as much as 20 cm/yr (Chen et al., 1993). The Devonian is also the time when land plants became widespread, influencing chemical weathering and erosion and carbon dioxide fixation into soils and coals. The plants began to decrease the proportion of carbon dioxide and to increase the proportion of oxygen in the atmosphere, thereby fostering cooling (Johnson et al., 1985; Geldsetzer et al., 1987; Gensel and Andrews, 1987; Buggisch, 1991; Shear, 1991; Wang et al., 1991, 1994, 1996; Guoqiu Gao, 1993; Joachimski and Buggisch, 1993; Algeo et al., 1995; Berner, 1997; Kenrick and Crane, 1997; Retallack, 1997). It seems likely that a mix of biogeochemical and geotectonic events during the Late Devonian–early Carboniferous transition brought about quick climate changes, including the waxing and waning of glaciers. Open water was close at hand and available for evaporation. Elsewhere over the Earth, and perhaps at great distance from the glaciers themselves, anoxic-water overturns, sea-level fluctuations, bolide impacts from time to time, and perhaps other events, may have also played a part.

ORDOVICIAN-SILURIAN ICE AGES

Short but severe episodes of glaciation occurred in Late Ordovician and Early Silurian time when advances of glaciers ended a long warm interval lasting for ca. 75 m.y. since the previous ice age: the Late Proterozoic Ice Age that ended in the Early Cambrian (Harland, 1972; Hambrey, 1985; Frakes et al., 1992; Eyles, 1993). The Ordovician-Silurian times of iciness took place at intervals during ca. 16 m.y. between ca. 445 Ma within the Late Ordovician Ashgill epoch until ca. 429 Ma at the end of the Silurian Llandovery epoch and may have lasted into early Wenlock time (Suárez-Soruco, 1995). These episodes of iciness are especially significant in understanding complex interrelationships bringing about glaciations because a quite complete worldwide stratal record is preserved, and during a time when biostratigraphic ranges of diagnostic fossils, especially graptolites and conodonts, allow reasonably satisfactory correlations and datings (Berry and Boucot, 1972, 1973a,b; Ross and Ross, 1992; Brenchley et al., 1994; Cooper et al., 1995). The interval is also marked by sea-level changes (Fig. 45) (Ross and Ross, 1985; Johnson, 1996) and by faunal extinctions and recoveries. Vascular plants did not yet flourish upon land to affect weathering

(Keller and Wood, 1993; Wyatt, 1995). Changes in carbon and oxygen isotopic ratios reveal geochemical variations in the world ocean (Marshall and Middleton, 1990; Berner, 1991; Brenchley et al., 1994; Gibbs et al., 1997; Finney et al., 1999). These changes suggest that world climate was near a balance and was triggered in and out of glaciation, at least partly by variations in greenhouse gases. It was a time of tectonic movements and Gondwana occupied a position near the South Pole. In the midst of this 16 m.y. interval of intermittent glacial advances, an especially sharp and brief episode of marked glaciation (perhaps less than 1 m.y. in duration) occurred in Hirnantian time (ca. 439 Ma) near the end of the Ordovician period.

In the Saharan region of *northern Africa* an indisputable Ordovician-Silurian glacial record is preserved. Here glacial landforms mantled by well-preserved continental glacial deposits are widespread. They are nicely illustrated in Beuf et al. (1971). The excellent geomorphic and stratal record is reminiscent of the Pleistocene record in North America! As shown by sculptured glacial features carved and scratched into the basement, and by tillites in overlying strata, major ice centers were sited upon the north African craton (Bennacef et al., 1971; Trompette, 1973; Deynoux, 1980, 1985; Biju-Duval et al., 1981; Deynoux and Trompette, 1981). The glacial strata are placed in the Tichit Group within the Taoudeni basin of west Africa, and a large ice cap to the south and southeast is inferred from widespread striations and directional-current structures within sandy interbeds (Deynoux, 1980, 1985; Deynoux and Trompette, 1981). Farther east in the El Hoggar region of north Africa, the glacial strata are assigned to the Tamadjert Formation (Beuf et al., 1971). In northern Africa the age of the glaciation is placed during the Hirnantian epoch (latest Ordovician) (Underwood et al., 1998), but may have lasted into the earliest Llandovery at the very beginning of the Silurian (Biju-Duval et al., 1981; Legrand, 1995; Paris et al., 1995). The dating depends on trilobites, graptolites, chitinozoans, and brachiopods recovered from marine intercalations. Llandoverian graptolites immediately and disconformably overlie glacigene beds in Arabia

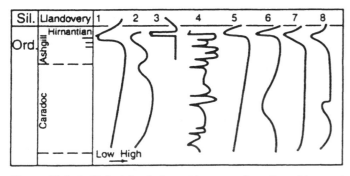

Figure 45. Late Ordovician bathymetric curves from Brenchley et al. (1994). Generalized curves are after (1) Legget et al. (1981), (2) Fortey (1984), (3) Brenchley and Newall (1980), and (4) Ross and Ross (1992). Regional curves are for (5) Yukon (Lenz, 1982), (6) central McKenzie Mountains (Lenz, 1982), (7) Poland (McKerrow, 1979), and (8) central Wales (McKerrow, 1979). Ord. = Ordovician; Sil. = Silurian.

(Vaslet, 1990). The uplifted cratonic areas from which ice flowed, now oriented northward and northwestward, are located in central Saharan Africa from the Moroccan region eastward through the Hogar and Libya (perhaps Egypt and Ethiopia) to Jordan and the Arabian Peninsula (Deynoux, 1985; McClure, 1978; Vaslet, 1990; McGillivray and Husseini, 1992; Abed et al., 1993). (Strata in Sierra Leone and Ghana, formerly considered tentatively as Ordovician in age, are probably Neoproterozoic [Reid and Tucker, 1972; Tucker and Reid, 1973, 1981; Deynoux, 1985]). Upland areas of northern Africa and Arabia harbored ice over an expanse some 3,500 km wide, but the glaciers are not known to be contiguous. Vaslet (1990, fig. 11; modified in Eyles, 1993, fig. 15.3) depicts an ice cap over central Africa, adjacent to the South Pole at about 430 Ma whereas in my review of both climatological and paleomagnetic data I elect to place the pole over South America at that time, as shown here on Figure 40.

In *southern Africa* a convincing record of glaciation is preserved in latest Ordovician strata of the Pakhuis Formation within the Cape Ranges (Rogers, 1902; Rust and Theron, 1964; Rust, 1975, 1981; Tankard et al., 1982). The age of the glacigene layers, based on fossils stratigraphically above (within the Cedarberg Shale) and below them, is considered as latest Caradocian (ca. 447 Ma) and early Ashgillian (ca. 444 Ma) (Theron, 1994). In the type area near Pakhuis Pass about 200 km north of Cape Town, the formation contains many faceted and striated stones within diamictite layers overlying a furrowed substrate, interpreted as formed when glaciers advanced into a sea across sediment layers that were firm but not lithified (Figs. 46, 47). The furrows are aligned with the same downflow directions as indicated by crossbeds in sandstones below, and upon flat surfaces on top of ridgelets between furrows that are at places glacially striated in the same direction (personal observations, July 1994). It seems likely that ice overrode outwash so that glacial scratches follow the same flow direction. Perhaps an ice lip rose and fell along with sea level, scribing the tops of furrows when it fell to bring moving ice in contact with the substrate. It is not known, however, whether such hypothesized rises and falls were short term, such as tidal, or long term such as the result of eustatic changes, including those resulting from glacier waxings and wanings. At least two distinctly different advances are recorded by tillites at the Pakhuis locality, with a marked break between them, but data are lacking to estimate the duration of the time break. From the type area of the Pakhuis Formation, intermittent outcrops with a glacial imprint extend southward for ~250 km (Söhnge, 1984). During Late Ordovician–Early Silurian time, the region was relatively near the margin of the Paraná basin in coastal Brazil, long before the South Atlantic Ocean opened.

In *South America* a glacial record is preserved in Brazil and Bolivia (Schlagintweit, 1943; Maack, 1957; Lohmann, 1965; Crowell et al., 1980; Rocha-Campos, 1981; Hambrey, 1985; Grahn and Caputo, 1992, 1994; M. V. Caputo, 1998, written communication). In Brazil, lobes of diamictite, viewed as deposited largely by ice rafting, extended into the Amazonas, Parnaíba, and Paraná basins from bordering highlands. Angular and subangular

Figure 46. View of glacially furrowed and striated pavement on consolidated Ordovician (Ashgillian stage) sandstone, Table Mountain Group, Pakhuis Formation. Cape Ranges, northeast of Clanwilliam at Pakhuis Pass, South Africa. August 1982.

Figure 47. Close-up of furrows at Pakhuis Pass locality in Figure 46.

basement clasts up to 1 m in size in the diamictites have been carried as much as 150 km from present basement outcrops along a depositional surface inclined gently at less than 1° in the Amazonas basin. At basin margins, underlying facies include fluviatile and littoral sediments in continental environments extending downflow into marine facies. Although striated dropstones in laminated deep-water beds have not been described from the Amazonian basins, facies relations suggest a glacial environment in part. The sequences indicate at least four advances and retreats of glacial lobes, all of probable Early Silurian age and assigned to the Llandovery stage (Grahn and Caputo, 1992, 1994). Diamictites cored in wells drilled for oil within the Paraná basin are probably glacigene and somewhat older and perhaps belong to the Ashgillian stage. Chitinozoans in the Amazonian basins are Llandoverian in age (Paris et al., 1995). Ice centers are interpreted as sited upon both the northern and southern margins of the Amazonas basin, to the west of the Parnaíba basin, and along the northern and eastern margins of the Paraná basin (Grahn and Caputo, 1992, 1994).

In *Bolivia*, and extending into adjacent Peru and Argentina, diamictites of the Cancañiri Formation contain rare faceted and striated boulders. These are interpreted as of glacial origin derived from uplands rimming an intermontane basin, perhaps a rift (Crowell et al., 1980). The youngest well-dated beds underlying the formation are of Caradocian age based on graptolites, trilobites, and brachiopods. From the glacigene beds, acritarchs and chitinozoans indicate a Wenlockian age (*Duvernaysphaera jelinii* zone), so the formation is given a latest Llandovery-earliest Wenlock age (Suárez-Soruco, 1995). Fossils of Ashgillian age have not been reported. Beds overlying the formation are placed in the *Neoveryhachium carminae* and *Pristiograptus colonus* zones of the Ludlovian stage in the Upper Silurian. In the absence of better dating, the Cancañiri Formation is considered primarily of Llandovery age but younging into the earliest Wenlock. In the cordillera of northern Argentina, Erdtmann et al. (1995) show the formation as unconformably overlying Ashgillian beds and assign it to the Llandoverian stage; Gagnier et al. (1996) place it in the Ashgillian.

The diamictite layers are associated with roiled beds including deformed sandstone bodies indicating deposition on or near a slope. They are therefore interpreted as having moved into the marine basin from highlands around the basin, but data are lacking to confirm a widespread ice cap. Tectonic reconstructions of northwestern Argentina and adjacent regions disclose high mountains in Pampeanas Ranges, capped with glaciers, which shed debris westward into a peripheral foreland basin (Astini et al., 1995). Along this belt basins were sinking with rising regions between them so that sea-level variations due to glacial waxing and waning cannot yet be separated from those due to tectonism (Crowley and Baum, 1991a; Keller, 1995). Several geologists have suggested that this Precordilleran belt is the far offset counterpart or conjugate of the eastern Laurentian belt in eastern North America (Ramos et al., 1986; Dalziel, 1991; Moores, 1991; Dalla Salda et al., 1992; Dalziel et al., 1994). It is therefore unlikely that the Cancañiri beds were laid down as close to the Brazilian basins along the Amazon trend, as they are now positioned, or near the Chiquerío Formation on coastal Peru.

Along the coast of southwestern *Peru* metamorphosed strata include boulder beds interlayered with mudstone, diamictite, and sandstone, and contain blocks up to 2–3 m in diameter (Caldas, 1979; Shackleton et al., 1979; Cobbing, 1981). I examined coastal exposures during a two-day excursion in September 1982, accompanied by Marco Fernandez-Dávila and Hugo Valdivia. At places in the sequence, assigned to the Chiquerío Formation (formerly Marcon Formation), large lonestone blocks penetrate laminations beneath them. Some exposures display small-scale sharp folds suggesting soft-sediment deformation, but now crosscut by metamorphic cleavage. We found no glacial polish or striations or facets on any of the megaclasts, but, if ever present, the metamorphism would be expected to obliterate them. We inferred tentatively that the boulder beds were probably laid down by debris flows and that the boulders with penetrations into the substrata might just as well have been deposited by rolling from a debris flow as by falling from icebergs. Perhaps the debris came from a nearby fault scarp. Glaciation was not necessarily indicated, although possible. Overprinting cleavage and small-scale structural crumpling interfered with sedimentological interpretations.

The large blocks appeared similar to pink granites and gneisses in the Proterozoic(?) basement of the Arequipa region, unconformably underlying the sequence, a sequence that is crosscut by gabbroic to granodioritic plutons dated by Rb-Sr methods at 392 ± 22 Ma (Shackleton et al., 1979). Related plutons in the region using U/Pb zircon methods reveal two magmatic events at ca. 425 Ma and 394–388 Ma (Mukasa and Henry, 1990). The Chiquerío Formation is therefore considered to have been laid down some time no later than an Ordovician-Devonian interval, but it might have been deposited during earlier times, including the Late Proterozoic, inasmuch as the age of the basement below the sequence is as yet unknown. The region, including these strata, requires modern investigation before we understand whether they document glaciation and, if so, at what time. The formation occurs within tectonic slices and may be far out of place, and not related to the Cancañiri Formation of Bolivia.

In western *Europe and North America* cold climates occurred at about the same time as the Gondwanan glaciations (Hambrey, 1985). Studies by Spjeldnæs (1961, 1981), based largely upon interpretations of fossil ecology and facies reconstructions, show that cool climates prevailed. In eastern Germany, the Lederschiefer of the Thüringer Schiefergebirge contains slightly metamorphosed diamictite layers with boulders up to 40 cm and some striated pebbles and a few places with possible dropstones (Steiner and Falk, 1981). Its stratigraphic position between fossiliferous beds both above and below indicates that the Lederschiefer is latest Ordovician or earliest Silurian in age. Glaciers in low upland regions to the south and east are viewed as shedding debris into an elongated basin. Stratal units elsewhere in Europe with features suggesting derivation in part from glaciers, but requiring much further investigation, are reported from

Normandy, France (Doré, 1981; Doré et al., 1985; Robardet and Doré, 1988; but Long, 1991, advocates a nonglacial origin), northeastern Scotland (Hambrey, 1985), Portugal and Spain (Arbey and Tamain, 1971; Fortuin, 1984), and western Ireland (Williams, 1980). In North America it is also not yet known whether interpretations of glacial contributions are warranted from strata in Newfoundland (McCann and Kennedy, 1974; Pickerell et al., 1979) or in Nova Scotia (Schenk, 1972). If so, originally tillitic debris may have been intermixed with other material moving downslope. Discussions of correlations in the cordillera of northern Canada involve the use of sharp sea-level changes that may have resulted from distant glaciation (Lenz, 1976, 1982; Johnson and Potter, 1976; Miller, 1976).

In summary, intermittent short episodes of cool climate over ca. 16 m.y. accompanied by both widespread and local glaciers are now well established during the time transition from the Ordovician into the Silurian. Data suggest that there were especially strong glacial waxings among several during this 16 m.y. interval: an earlier one in Hirnantian time and two or three in Llandovery time. An associated marked drop in sea level between 50 m and 100 m is inferred from stratal sections in several regions in late Ashgill (Hirnantian) time (Woodcock and Smallwood, 1987; Eyles, 1993; Ross and Ross, 1988; Heredia and Beresi, 1995) and perhaps as much as 150 m and similar to sea-level changes in the late Cenozoic (Wyatt, 1995). The sea-level drop was accompanied by sharp excursions in both oxygen and carbon isotopic ratios, particularly during the Hirnantian (Brenchley et al., 1994; Kump et al., 1995). Biotas were severely affected by this short-lived time of coolness and related events so that major extinctions and reradiations occurred (Sheehan, 1973). It seems likely that warm, saline bottom waters of the ocean changed suddenly into a system characterized by cold deep-water circulation and then back again (Railsback et al., 1990; Brenchley et al., 1994). Although theoretical and modeling studies suggest a warm Earth and high values of greenhouse gases in the atmosphere and oceans during the early Paleozoic, coolness occurred at times during this interval when ice caps grew in near-polar regions (Berner, 1987, 1990; Crowley and Baum, 1991b, 1995; Kasting, 1992a; Frakes et al., 1992). I view the basic cause of these dramatic changes in both climate and extinctions as rooted in the movement of continental cratons with respect to the South Pole and in rapidly changing arrangements of continental blocks as they actively glided about in southern near-polar regions (Scotese, 1990; Van der Voo, 1993; MacNiocaill et al., 1997). Ocean bathymetry and continental arrangement affected ocean-current circulations including deep-water overturnings and flooding of continental shelves and margins by nearly anoxic waters. The carbon dioxide content of the atmosphere may have been near a threshold so that these events dramatically affected climate. It may also have been a time of quite rapid continental movement as inferred from some paleomagnetic studies showing a loop across South America and back during the Silurian (Kent and Van der Voo, 1990). Further investigations of paleogeographic and paleobathymetric arrangements and of paleomagnetism during this critical interval are required.

Although the sharpest change in environmental conditions for the Ordovician-Silurian ice ages took place in Hirnantian time near the end of the Ordovician, the stratigraphic record as outlined above shows that episodes of iciness probably began in the Caradocian and certainly extended well into the Silurian. In Brazil, for example, at least three distinct levels of glacially derived sediments are placed within the Llandoverian stage (M. V. Caputo, 1998, written communication). The strongest glaciation, in the Hirnantian, occurred when the northwestern African part of Gondwana lay near the pole. Later, in the Early Silurian, as the cratonic blocks moved, Brazil and Bolivia were arranged so that ice grew in these regions. In addition, basins in South America were beginning to develop during Late Ordovician time suggesting that highlands at basin margins during the earliest Silurian may have provided sites to augment glacial growth. The Silurian glaciations may therefore have continued in mountains rimming depositional sites following upon the main Hirnantian episode. Attractive explanations for the Ordovician-Silurian glaciations therefore include rapid tectonic movement of cratonic blocks accompanied by arrangements favoring strong deep-water flow and upwelling. With glaciation, marked sea-level drops occurred, followed by rapid rises as the glaciers dwindled, and were accompanied by critical changes in water vapor and carbon dioxide flux and chemical weathering. In the early Silurian, mountains grew marginal to basins, perhaps including rifts, and glaciers grew upon them.

LATE PROTEROZOIC ICE AGES

Glaciations during Late Proterozoic time (Neoproterozoic) are recorded at places on all continents if we include the probable record in Antarctica. Moreover, interpretations of the geologic record of the Neoproterozoic suggest that the mix of processes influencing climate were at times different from those that prevailed during the Phanerozoic. The icy episodes occur back from Early Cambrian time (ca. 520–530 Ma) to perhaps ca. 950 Ma, an interval of some 430 m.y., as long as the time span from the present back into the Silurian (Chumakov, 1981; Crowell, 1981; Trompette, 1982, 1996; Harland, 1983; Hambrey and Harland, 1985). For the ca. 75 m.y. following the Late Proterozoic Ice Age until the Late Ordovician glaciations at ca. 445 Ma, there is no confirmed record of glaciation. Climate on Earth warmed.

Several glacial episodes are documented during Neoproterozoic time but as yet are poorly dated. From youngest to oldest they include: Glaciations younger than the first appearance of Ediacaran fossils, between 545 Ma and 585 Ma (perhaps as young as 520 Ma in the Early Cambrian); those between 590 Ma and 640 Ma, termed Marinoan or Varangian by many investigators; those between 700 Ma and 750 Ma (commonly labeled Sturtian), and those ca. 900 Ma in age with a large uncertainty.

The dating of geological and climatic events in the Proterozoic is quite insecure compared with dating in the Phanerozoic, except at a few places such as for the Nama Group in Namibia

(Saylor et al., 1995, 1998; Kaufman et al., 1997). This is because of the lack of volcanic and other rock units giving isotopic ages in appropriate geologic settings and the lack of time-constrained fossil ranges. During the next few decades, however, dating can be expected to improve as new localities are discovered where radiometric methods can be applied, and where there is improvement in the correlation potential of distinctive isotopic excursions and sea-level changes and in the knowledge of Proterozoic fossil ranges. At present, however, in moving back into the Proterozoic, our judgments in regard to ages are relatively infirm on a worldwide basis. Only locally is the dating satisfactory. Moreover, with increasing antiquity, tectonic reconstructions become progressively more and more ambiguous and hazy, and time slices are longer and far less synchronous.

In the history of ancient paleoclimate research, Late Proterozoic glaciations have drawn special attention. Some of the problems raised are now partly resolved but others remain. As paleomagnetic data accumulated beginning about four decades ago, some localities with upper Proterozoic strata displayed both a glacial record and a low latitude. This led to the concept of a worldwide glacial interval, the great "infra-Cambrian ice age," when true glacial ice was viewed as extending across the Earth from polar into equatorial regions (Harland and Bidgood, 1959; Harland, 1964; Tarling, 1974; Chumakov and Elston, 1989). For several decades, many geologists (including me) were skeptical of the paleomagnetic measurements and were reluctant to build climatic hypotheses on their basis. However, paleomagnetic measurements and interpretations of some Neoproterozoic glacigene beds in Australia and southern Africa now do indeed support the concept of low-latitude iciness (Embleton and Williams, 1986; Schmidt et al., 1991; Kirschvink, 1992; Powell et al., 1994; Meert and Van der Voo, 1994, 1996; Schmidt and G. E. Williams, 1995; Meert et al., 1995; G. E. Williams et al., 1995; Sohl and Christie-Blick, 1995; Kirschvink et al., 1997; Sohl, 1997; D. M. Williams et al., 1998). What is the explanation of this apparent anomaly that suggests that climate belts in Neoproterozoic time were different from those today when glaciers occupy polar regions and equatorial regions are relatively ice free?

Discussion and controversy have also involved the interpretation of several Proterozoic glacigene sequences that are capped, interlayered with, or underlain by carbonate beds. Many of these are interpreted as requiring warm or subtropical waters, deemed characteristic of low latitudes. However, some carbonate facies are laid down in cool or cold water and may be largely made up of detrital carbonate grains eroded from carbonate-rich substrata (Fairchild and Spiro, 1990; Fairchild 1993; Fairchild et al., 1994). They cannot necessarily be interpreted as indicating warm water associated with tropical oceans. Other carbonate beds have been deposited well below high mountains with slopes steep enough to allow glacial material to intermix with debris flows and so connect alpine glaciers with warm-water basins. In addition, the rise of sea level across flat coastal plains following glaciation may bring about a wide belt of very shallow water easily warmed by the sun; an increase in temperature reduces the solubility of carbonates. Such an environment is propitious for carbonate-utilizing creatures, and as the ambient temperature warms, to calcium carbonate precipitation. Biogenic carbonate layers overlying Pleistocene tillites are viewed as forming at present through this chain of processes on shelves bordering the Barents Sea, northern Norway (Bjørlykke et al., 1978). Recently Hoffman et al. (1998a,b) suggest that earthwide iciness inhibited the exchange of carbon dioxide between the seas and the air until there was postglacial buildup of carbon dioxide from volcanic activity, accompanied by rapid warming. This brought about the precipitation of cap carbonates. In short, where capping or interbedded carbonates occur with glacigenic sediments, both a local and regional reconstruction may be required to fix their origin. Such carbonate layers therefore require pointed attention.

The investigation of isotopic excursions preserved in critical stratigraphic sections is especially important for Neoproterozoic sequences (Fig. 48). Studies of geochemical trends in Neoproterozoic carbonates, especially those of C, S, and Sr isotopes associated with glacigene intervals, are revealing strong excursions indicating marked changes in oceanographic and climate regimes (Kaufman and Knoll, 1995; Kennedy, 1996; Kennedy et al., 1998). Documented excursions, however, exceed in magnitude those recognized in Phanerozoic strata and suggest global events. They have resulted from biogeochemical processes during marked evolutionary changes leading into the burgeoning of biota in the Phanerozoic. The samples examined have passed tests to assure that the ratios are primary and not altered significantly by diagenesis or metamorphism. Such chemical variations require calibration with the time scale that is now being achieved by dovetailing with radiometric dates, sea-level curves, sequence stratigraphic units, and biostratigraphic data (Derry et al., 1989, 1992; Asmeron et al., 1991; Des Marais et al., 1992; Knoll et al., 1986, 1992; Kirschvink, 1992; Knoll and Walter, 1992, 1995; Kaufman and Knoll, 1995; Knoll et al., 1996; Kaufman et al., 1997; Hoffman et al., 1998a,b; Kennedy, 1996; Kennedy et al., 1998). Unfortunately, paleogeographic reconstructions are not yet good enough to judge water depth, size of basins, irregularity of bathymetry, and other factors needed to evaluate ocean-water stratification and the flow of bottom waters, including their occasional overturn.

Several hypotheses are under consideration now to explain the Proterozoic low-latitude glaciations: (1) those based upon a climate system controlled by processes quite similar to those operating during the Phanerozoic, (2) those calling upon a "snowball Earth" with ice extending from high to low latitudes, (3) those proposing a greater inclination of Earth's spin axis with respect to the plane of the ecliptic, (4) those based upon the possibility of "true polar wander," and (5) those involving extraterrestrial circumstances and influences.

Explanations of low-latitude glaciation rooted in the concept that the tectonic world operated much as it does today call upon the mobility of crustal blocks and changing continental rearrange-

ments to explain climate change. The view is that these mobilities were dependent upon the internal thermodynamics of the Earth and that plate tectonics, or a process quite similar, prevailed then as now. Were crustal blocks markedly mobile so that glacial sites moved rapidly into low-latitude sites from middle latitudes where glaciation is known to have occurred in the Phanerozoic? Because of rapid plate-tectonic movements, was the magnetic imprint fixed into the strata after the tectonic block had moved out of subpolar or middle-latitude regions? Were the continents clustered in equatorial regions so that deep tropical weathering and CO_2 drawdown were enhanced to the extent that an icy threshold was crossed (Kirschvink, 1992)? Was there an abrupt overturn of oceanic waters at the time of the Cambrian-Neoproterozoic transition (Kimura et al., 1997)? Were glaciers absent from polar regions because there were no polar continental sites? Were continental topographies and oceanic bathymetries so different or so

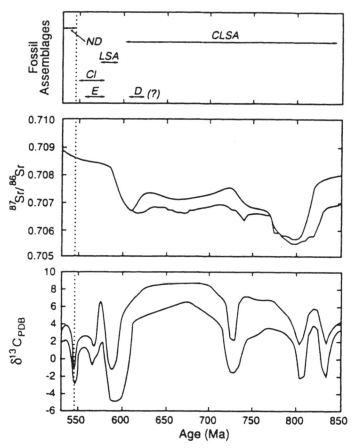

Figure 48. Estimated ranges of fossil assemblages and variations in $^{87}Sr/^{86}Sr$ and $\delta^{13}C$ in Neoproterozoic to basal Cambrian carbonates (ca. 850–530 Ma). Refer to Kaufman and Knoll (1995) for fossil-bearing localities and their correlations: ND = Nemakit-Daldyn small shelly fossils; CLSA = Large acanthamorphic acritarchs of Cryogenian aspect; LSA = Large spiny acritarchs of Terminal Neoproterozoic III aspect; CI = *Cloudina*; E = Diverse Ediacaran faunas; D = Disc-shaped fossils of possible metazoan origin. PDB = Peedee Formation belemnite, Cretaceous, South Carolina, used as a standard. Simplified from Kaufman and Knoll (1995, fig. 6).

arranged that climate was also markedly different? Were biochemical products of Neoproterozoic life, through effects on the carbon and sulfur cycles, also different from today? Were many of the low-latitude sites next to high mountains, such as shoulders of rifts or along continental arcs, so that high alpine glaciers shed their debris directly into nearby low basins (Schermerhorn, 1974; Bjørlykke, 1985; Eyles, 1993; Eyles and Young, 1994; Trompette, 1994, 1996)? Was glacial debris carried by icebergs in ocean currents for great distances into low latitudes from glaciers that reached the sea in middle latitudes before dumping their telltale debris upon a basin floor (Crowell, 1964)? Unfortunately, the Neoproterozoic glacial record as known today provides only partial answers to these and similar questions.

The "snowball Earth" hypothesis pictures a time of iciness extending from poles to low latitudes so that the seas and oceans are covered with ice as well as the continents (Kirshvink, 1992; Hoffman et al., 1998a,b). The hypothesis follows as a modification from the concept of the "Great Infra-Cambrian Ice Age" (Harland, 1964). On land a thin blanket of ice is proposed but thick ice is as unlikely. According to this concept, isotopic excursions and carbonate layers overlying glacigene sequences (the cap carbonate) are interpreted as resulting from earthwide thin ice that blocked moisture interchange between the seas and atmosphere. This situation was followed by rapid warming as volcanic activity added greenhouse gases to the air (Hoffman et al., 1998a,b). As a consequence, the thin and widespread ice rapidly melted so that the cap carbonate was deposited in warming shallow water, especially upon broad continental shelves. However, an explanation is still needed for the glaciation itself, which took place before the deposition of the cap carbonate.

An explanation for the low-latitude glaciation involving a difference in the orientation of the Earth's spin axis with respect to the plane of the ecliptic has been proposed and defended since the early 1970s by G. E. Williams (1975a,b; 1993, 1994; Schmidt and G. E. Williams, 1995; D. M. Williams et al., 1998; Rubincam, 1995). During the past quarter century, many geologists (including me) have dismissed such an explanation because they saw little reason to abandon a uniformitarian approach to ancient Earth history and were skeptical of the paleomagnetic data upon which the hypothesis rested. We preferred an explanation that required only a slight departure from a conventional uniformitarian approach. Williams proposed that the Earth's rotational axis had been tipped with respect to the plane of the ecliptic to an angle of 54° or greater. According to this arrangement, the polar axis would point much more toward the sun than at present so that low latitude regions were relatively shielded from the sun's heat flux and cool enough to allow glacier growth on continents near the equator. Polar regions, in contrast, would receive strong solar radiation with the North and South Poles pointing sunward alternately according to the season. Tidal friction and other factors affecting the obliquity are under consideration (Ito et al., 1995; Jenkins and Frakes, 1998).

True polar wander envisions the sliding of outer shells of the Earth upon discontinuities within or at the base of the mantle (e.g., Piper, 1987; Kirschvink et al., 1997). Sliding of these shells transports both the upper mantle and lithosphere and carries with them plate-tectonic movements and displacements. Apparent polar wander paths plotted from paleomagnetic measurements are the result of at least three processes rooted within the internal thermodynamics of the Earth: those resulting from plate-tectonic movements, those related to movements across "hot spots" rising from deep with the mantle (Morgan, 1981), and those from "inertial interchange true polar wander," usually referred to as "true polar wander." Kirschvink and colleagues (1997) suggest that an unstable distribution of mass within the lithosphere and upper mantle around the spin axis resulted in rapid continental movements during Early Cambrian time. It was a response to the planet's moment of inertia and resulted in a quick transition to a longitudinal Cambrian paleogeography. They relate these changes in surface conditions to the marked diversification of life and its rapid evolution. If this explanation is proven, similar rapid paleogeographic rearrangements severely affecting the climate system may well have occurred earlier in Precambrian time, but as yet there is no reported evidence for such events. If an event is documented at one time in Earth history, it is prudent to search for evidence documenting similar events back within the dim vistas of ancient time.

Extraterrestrial or astronomic events and processes, in addition to those associated with orbital arrangements, are also under consideration as affecting the climate system in the remote past. First, perhaps the system was influenced significantly by bolide impacts (e.g., Alvarez, 1997). Second, galactic events and, third, the evolution of the Earth-moon system have been proposed as influencing Precambrian climates (e.g., Steiner and Grillmair, 1973; Salop, 1977; Pollack, 1982; Williams et al., 1998). Several theoretical treatments suggest that the Earth-moon distance and the length of the day have increased since the Proterozoic but that the system has been mainly stable since the moon's capture by the Earth very early in the history of the solar system (Sonett et al., 1988; Berger et al., 1992; Taylor, 1992, 1998; Laskar and Robutel, 1993; Sonett et al., 1996). On the other hand, Williams et al. (1998) present arguments to show that the history of the moon's orbit has changed through time. Perhaps the irregularities in the moon's orbit today reveal that they remain from a time when the angle between the Earth's spin axis and the plane of the ecliptic has diminished from a Precambrian value of something around 54° to its present value of 23.5°. Williams et al. (1998) also propose that the high obliquity was the result of imbalance brought about by continental arrangements and associated large ice sheets. Muller and MacDonald (1995, 1997) present theoretical arguments that the planet's orbital plane had a different tilt in the remote past. In addition, Crowley and Baum (1993) find that their modeling studies, although based on crude paleogeographies, allow for glaciation in latitudes as low as 25° from the equator if the solar luminosity was about 6% less than at present (Sagan and Chyba, 1997; Kasting, 1992a,b; Kasting et al., 1992). Another hypothesis that has received little support is that the Earth might have possessed ice rings similar to the rings surrounding Saturn, and arranged so that they reduced insolation falling upon equatorial regions, thus enhancing cooling in low latitudes (Sheldon, 1984).

Any hypothesis to explain low-latitude glaciation requires consideration of such questions as: What would be the consequences and what would be preserved within the geologic record? For example, if there had been an especially severe worldwide glaciation sequestering much water within land glaciers and lasting for a long time, sea level would have been markedly lowered. This record ought to be preserved somewhere. A marked drop in sea level would be expected, but few have been noted. An example may consist of some deep, infilled submarine canyons in Australia (Christie-Blick et al., 1990b), but local tectonic uplift and subsidence of a rift margin may account for their cutting and filling. If continents were clustered in middle and lower latitudes so that polar regions were relatively ice free and glaciers were concentrated in middle and low latitudes, what would be the effect on the location of the belt of carbonate rocks that now circles the Earth in tropical and subtropical regions with a latitudinal arrangement? If more heat had arrived from the sun in polar regions, would the carbonate belt lie poleward? Do continental reconstructions and correlations show that there were both high- and low-latitude glaciations at the same time? Does the stacking arrangement of sun-seeking stromatolites, which seems to have the same orientation ca. 850 m.y. ago as today, suggest that the spin axis had nearly the same tilt (Vanyo and Awramik, 1982; Awramik and Vanyo, 1986)? Proterozoic continental reconstructions do not as yet allow worthwhile speculation on answers to these questions.

The Neoproterozoic icy record is next reviewed before approaching answers to some of these questions, but answers to most must await future research.

Africa

A Late Proterozoic glacial record, perhaps extending into the Early Cambrian, is well preserved in the northwestern *Saharan region* of Africa and along stretches of the irregular Pan-African foldbelt. The tectonostratigraphic regimes include cratonic sites and convergent regimes along continental margins and along complex mobile belts. Inland from the northwestern bulge of Africa there are undoubted glacial sediments preserved within the Taoudeni basin, assigned to the Jbéliat Group (Fig. 49) (Deynoux and Trompette, 1976, 1981; Trompette, 1981; Deynoux et al., 1978). Here glacial landforms and overlying sedimentary facies are reminiscent of the Pleistocene record. This broad cratonic basin, at least 1,500 km wide, derived debris from rising orogenic belts along its margins (Robineau and Ritz, 1990; Hefferan et al., 1992; Trompette, 1994). Nonglacial diamictites formed as the result of downslope sliding from rising mountains interfinger eastward into facies with a glacial imprint, suggesting the spread of piedmont glaciers at the edge of the wide basin. The Jbéliat glacial beds lie between Rb-Sr isotopic dates in clays of 595 Ma and 630 Ma (Clauer and Deynoux, 1987). Along the northern margin of the Taoudeni basin, tillites are reported in strata lying

above Late Proterozoic Ediacaran faunas, at the boundary with the Cambrian (Culver et al., 1988; Bertrand-Sarfati et al., 1995; Trompette, 1996). In the southern part of the Taoudeni basin glacigene beds thicken laterally into the Bakoye Group where several stacked sequences are probably related to sedimentation cycles and glacial advances. Although the timing of these glacial advances into the subsiding basin is uncertain, a total duration of ca. 10 m.y. is proposed (Deynoux et al., 1989, 1991; Proust and Deynoux, 1994). Probable glacial marine deposits also flank the central Saharan Touareg Shield of the Hoggar region in Algeria and Mali (Caby and Fabre, 1981a,b; Deynoux, 1985; Trompette, 1996). Here lonestones interpreted as glacial dropstones occur in Cambrian carbonate strata and above a regionally correlated "triad" consisting of glacigene beds, cap carbonates, and surmounted by bedded cherts. This triad is correlated as the same transgressive unit across the west African craton to the Volta basin to the south (~1,200 km) and to the Hoggar region to the southeast (~1,500 km). These glacial deposits are as young as ca. 520 Ma (Early Cambrian) if correlations are correct.

Bordering the Saharan region on the south and west, glacial beds are described in Sierra Leone, Ghana, Senegal, and Guinea, and were previously considered as Ordovician in age (Reid and Tucker, 1972; Tucker and Reid, 1973, 1981). In Guinea and Senegal at the western edge of the west African craton, glacigene marine diamictites and laminated units with dropstones capped

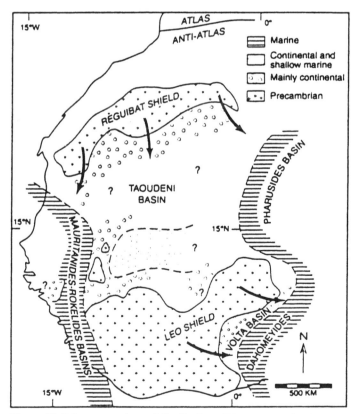

Figure 49. Paleogeography and distribution of Neoproterozoic glacigene facies across west Africa. After Deynoux (1985, fig. 2).

by dolomites are assigned to the Mali Group (Villeneuve, 1989). These glacial deposits are bracketed in age between 550 Ma and 660 Ma. In the Volta basin (mainly Ghana), terrestrial glacial deposits are flanked by lateral marine equivalents in the Dahomeyides orogenic belt (Trompette, 1981; 1996). The beds of the Volta basin are separated from those of the Taoudeni basin and its margins by the Leo Shield, which was a continental source for debris shed into the basin. Far to the east, upper Proterozoic beds with a possible glacigene aspect are also reported in equatorial Uganda (Bjørlykke, 1973).

In *western Africa south of the equator* (mainly Angola, Namibia, and South Africa), glacial units interfinger with thick deposits of other origins (Martin et al., 1985). These units occur within and along the irregular Pan-African rift system and mobile belt (Trompette, 1994; Prave, 1996). Deformed rocks now occur along both sides of the Atlantic Ocean, which opened in Mesozoic time and cut across older tectonic trends, such as those formed during and after Neoproterozoic time. Three cratons meet in this broad region: the Rio de la Plata craton in South America on the west, and the Congo and Kalahari on the south in southern Africa (Prave, 1996; Unrug, 1991, 1996, 1997; Pedrosa-Soares et al., 1998). These cratons finally amalgamated in Cambrian time to form part of Gondwana. Between the Congo and Kalahari cratons the Damara belt trends northeasterly and contains compressed and deformed rocks laid down in the Khomas Sea or trough, including part of the spread of the glacigene Chuos Formation (Henry et al. 1986; Badenhorst, 1988; Kukla and Stanistreet, 1991; Stanistreet et al., 1991). Rock units between the two cratons include remnants of an oceanic floor along with slices of a former accretionary prism, now caught within the collision belt between the two cratons (Kukla and Stanistreet, 1991; Frimmel et al., 1996). The width of the former sea and trough between the two cratons is therefore largely unknown. Interpretations suggest that when the Damara mobile belt was tectonically active, rift margins were raised and basins between blocks were depressed (Stern, 1994; Dürr and Dingeldey, 1996).

Glacigene strata lie upon the western portion of the Congo craton on the north and upon the Kalahari craton south of the Damara orogen. These strata include diamictites containing faceted and striated clasts, associated dropstone facies, and bedding surfaces with soft-sediment grooves interpreted to have been overridden by ice (Tankard et al., 1982; Kukla and Stanistreet, 1991). Structural and metamorphic overprinting, locally associated with datable plutons and volcanic rocks, provides clues to the tectonic and sedimentation history, but mapping and investigations in this large and complicated region are incomplete. Unfortunately, because of remagnetization, paleomagnetic data are not helpful in providing insight on the location of the two cratons before collision (Meert and Van der Voo, 1994; but see Williams et al., 1995). Fortunately, isotopic excursions provide clues to processes in the ancient oceans and may in time aid in correlation between far-separated sequences.

In *northern Namibia* upon the southwestern tip of the Congo craton, perhaps now constituting a separate Angola plate

(Frimmel et al., 1996), upper Neoproterozoic glacial beds and capping carbonates interbedded with shales have recently been investigated (Hoffman et al., 1998a,b; Kennedy et al., 1998). These beds are interpreted as laid down upon a huge low-lying platform as large as the conterminous United States. The glacigene units include the Chuos Formation, lower in the section, and the higher Ghaub Formation that consist of diamictites carrying outsized stones and locally laminated facies crowded with dropstones. Strong negative excursions up to -5 $\delta^{13}C$ ppm are associated with the Ghaub Formation and the overlying cap carbonates. Strontium ratios [$^{87}Sr/^{86}Sr$] as low as 0.7072 are obtained from the Ghaub as well (Kennedy et al., 1998). Distinctive "tubestones" consisting of millimeter-scale vertical structures characterize a marker bed in carbonates immediately underlying the stratigraphic level of the Ghaub where it is absent. In the absence of datable units within the immediate vicinity, the beds are correlated with sections at distance in southern Namibia where well-dated and isotopically analyzed rocks lie upon the Kalahari craton above glacigene sequences (Saylor et al., 1995; 1998). These dated beds occur on the other side of the Damara orogen, which here is about 400 km wide and was no doubt much wider before collision of the two cratons (Kukla and Stanistreet, 1991; Prave, 1996). The glacigene beds lying upon both cratons are given an age between 700 Ma and 760 Ma, although dating comes from the Kalahari region on the south and also from units stratigraphically overlying them. The glacigene Chuos Formation in northern Namibia is correlated with the Blaubeker tillite of the Khomas trough and with the Kaigas tillite of southern Namibia; the Ghaub probably correlates with the Numees Formation (Germs, 1995).

In *southern Namibia and adjoining South Africa*, the principal glacigene formations, lying upon and near the western margin of the Kalahari craton, are the Blaubeker and Bildah. Facies changes indicate that glaciers shed debris intermixed with other material westward onto slopes and shelves and into thick basinal sequences. The upper Neoproterozoic section overlying these glacigene units contains at least 11 ash beds amenable to radiometric dating along with associated carbonates for studies of variations in stable isotopic ratios (Saylor et al., 1995, 1998; Kaufman et al., 1997). The sequence includes beds revealing the first appearance of Ediacaran-type fossils, so this is one of the few places on Earth where significant biologic and other events are successfully dated and where stable isotopic curves are time calibrated. In the thick underlying sedimentary prism below the glacigene Numees and its correlatives, other probable glacigene diamictites have been recognized, including the Kaigas tillite. Diamictites lying well above the Numees, such as the Nomtsas, are not known to contain glacigene facies (Germs, 1995; Saylor et al., 1995, 1998). In this younger section, however, several unconformities and interpretations from sequence stratigraphic analysis indicate sharp sea-level changes. These sea-level fluctuations ended at about 550 Ma but had been underway for at least 200 m.y. previously (Tankard et al., 1982; Kaufman and Knoll, 1995). Note that this dated sequence, younger than glacigene beds lying upon the Kalahari craton, contains Ediacaran remains, and that glacigene sequences in northwestern Africa lie above Ediacara (Bertrand-Sarfati et al., 1995). In the belt between the Kalahari and Congo cratons, glacigene beds have been tectonically overprinted and even metamorphosed at places (Figs. 50, 51).

Rocks in the region of southern Africa have been studied paleomagnetically with results suggesting low paleolatitudes (Kröner et al., 1980; McWilliams, 1981; McWilliams and Kröner, 1981) but Van der Voo and Meert (1991; Meert and Van der Voo, 1994, 1995; but see Williams et al., 1995) conclude that they have been remagnetized. Moreover, they note that the position of the Kalahari craton during the interval from 900 Ma to 600 Ma, employed to produce an apparent wander path, is unconstrained paleomagnetically and that age assignments are quite approximate. Paleomagnetic measurements from the eastern part of the Congo craton, mainly from Neoproterozoic lavas in Kenya and Tanzania, give a paleolatitude between ~39° S at 547 ± 4 Ma and ~12° S at 743 ± 30 Ma (Meert et al., 1995; Meert and Van der Voo, 1996). They conclude that the paleomagnetic data do not require a latitude lower that ~25° S from sites well to the south and upon the margin of the Kalahari craton (Meert and Van der Voo, 1994; Van der Voo and Meert, 1991). Moreover, the southern part of the Congo craton may be tectonically separated from the distant eastern part in Kenya and Tanzania, constituting the separate Angola plate, so that the latitudinal position of the Congo cratonic region in northern Namibia is not definitely known (Frimmel et al., 1996). I therefore think it prudent to delay concluding that the Namibian glacial beds were deposited at low latitudes until critical paleomagnetic and dating studies are undertaken if appropriate sequences are present.

Cap carbonates directly overlie the Chuos and Ghaub glacial units in the northern Namibia study region (Hoffman et al, 1998a,b; Kennedy et al., 1998). Carbon isotopic investigations of these irregular thin-bedded dolomites, limestones, and shales show large negative ^{13}C excursions of as much as $-6‰$ from a preglacial value of from +5 to +9. Hoffman and colleagues (1998a,b) conclude from sequence stratigraphic and subsidence analyses that deposition upon a broad and gently sloping platform took place over ca. 10 m.y., an estimate including the time of deposition of both the glacial units and the cap carbonates, but the latter would be laid down quite rapidly. As the sea retreated across the continental shelf, the ^{13}C values decreased and then increased as the sea advanced again and the sea temperature warmed. As discussed in the summary below, they view these events as worldwide and mainly the consequence of Neoproterozoic glaciation.

Within *south-central Africa*, mainly in Zambia and Zaire, and also along the belts of deformed rocks and shear zones between the Congo and Kalahari cratons extending eastward from the main Pan-African belt (Stern, 1994; Unrug, 1996, 1997), there are diamictites interbedded within poorly dated Proterozoic strata (Alvarez and Maurin, 1991; Eyles, 1993; Trompette, 1994). Here, both an upper and lower "tillite" or conglomeratic sequence is interpreted as derived in part from glacial sources at basin mar-

Figure 50. Neoproterozoic Chuos pebbly schists, Naos region, southwest of Windhoek, Namibia, interpreted as of partial glacigene origin. Spine of notebook is 18 cm long. August 1982.

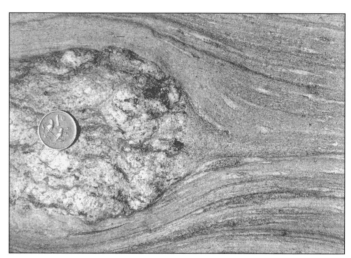

Figure 51. Cobble in Chuos deformed metadiamictite occurring in schistose-layered metapellite. Naos region, southwest of Windhoek, Namibia. Coin is ~2 cm in diameter. August 1982.

gins (Cahen and Lepersonne, 1981a,b; see also Schermerhorn and Stanton, 1963; Binda and Van Eden, 1972; Schermerhorn, 1974; Deynoux et al., 1978). The dating of the lower conglomeratic sequences is placed at ca. 900–950 Ma (Trompette, 1994), and regional arguments indicate that it is significantly older than later Proterozoic glacigene strata dated between ca. 700 Ma and ca. 760 Ma. Zircon U-Pb isotopic dates of a syntectonic batholith of ca. 820 Ma indicate a long and complex history (Stern 1994) although the age of the older conglomeratic sequences is as yet assigned only to the Late Proterozoic by Hanson et al. (1988). These beds, probably having a partly glacial source, are correlated by Trompette (1994) with rift deposits of Brazil where glacigene deposits are found within the Araçuai fold belt, the complex orogenic system at the western margin of the combined São Francisco–Congo craton. They cannot be correlated with deposits in northwestern Africa.

Australia

Several Neoproterozoic basins that preserve strata with a glacial imprint are located across the middle of Australia from the Kimberley region on the present northwest to Tasmania on the southeast (Figs. 52, 53). These basins contain at least two glacial sequences: the lower (Sturtian) and the upper (Marinoan). In addition, in the Kimberley region, there may be evidence of a third glaciation (Egan), either younger than Marinoan or a long-lasting part of the Marinoan (Plumb, 1993; Grey and Corkeron, 1998). A glacial contribution to these Australian sequences is quite certain although discussion prevails concerning their paleogeographic and tectonic setting, their age, and the depositional paleolatitude of some Marinoan sites as judged from paleomagnetic interpretations. Several sequences are over-

lain by cap carbonates with sharp negative $\delta^{13}C$ excursions (Williams, 1979; Kennedy, 1996).

The glacial interpretation is based upon the presence of glacially striated and sculptured basement outcrops, especially along the southern margin of the Kimberley block (Figs. 54, 55), and upon widespread units containing glacially faceted and striated clasts dispersed within a homogeneous muddy matrix, as well as dropstones occurring in laminated sequences. Strata revealing a glacial imprint are found in three settings: within the Adelaide "geosyncline" or trough or foldbelt. In my view, the term "geosyncline" should fall into disuse although it is ingrained in the literature and still widely used. It is a holdover from pre-plate tectonic concepts. I prefer "trough" when the emphasis is placed upon depositional sites and events, and "foldbelt" when the context is placed on deformational events subsequent to deposition and upon the belt as it occurs geographically today. Similar beds within remnant basins crossing central Australia may have originally been parts of a wide Centralian superbasin (Walter et al., 1995; Myers et al., 1996). Correlations from basin to basin

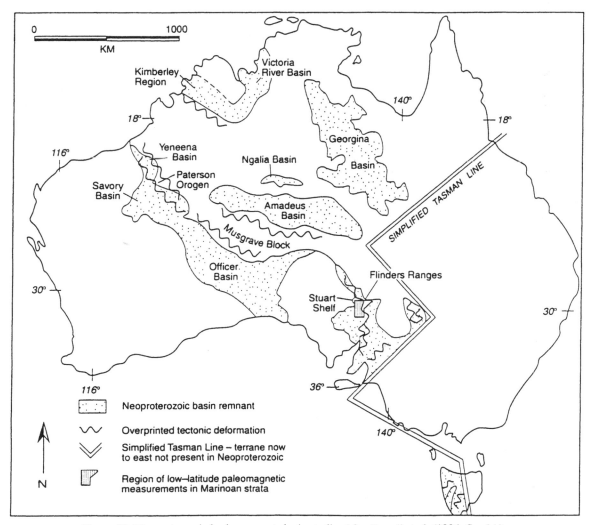

Figure 52. Neoproterozoic basin remnants in Australia. After Powell et al. (1994, fig. 1A).

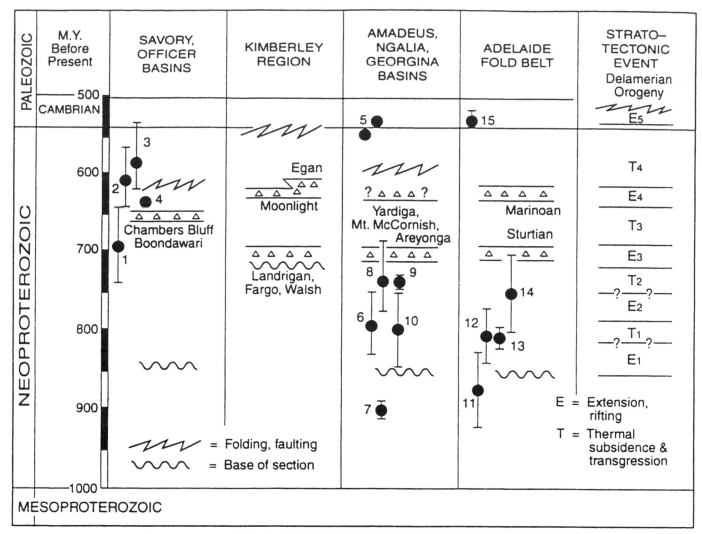

Figure 53. Neoproterozoic regions of Australia, showing dating, stratigraphic position of tillitic layers, and tectonic events. (1) Age of emplacement of Mount Crofton granite; (2) recalculated Rb-Sr age of Mount Crofton granite; (3) Rb-Sr age of Table Hill volcanics; (4) preliminary date for dolerite dikes intruding glacigene Boondawari Formation, Savory basin; (5) ^{40}Ar/^{39}Ar date on timing of thrusting; (6) Sm-Nd dates on various mafic dikes; (7) Rb-Sr date on dike swarm; (8) estimate of age of Pertatataka Formation, eastern Amadeus basin; (9) age of the Mud Tank cabonatite, Northern Territory; (10) from Powell et al. (1994, fig. 9); (11), (12), (13) Sm-Nd dates on various mafic dikes; (14) Pb-Pb age on zircons from lower Adelaidian tuff; (15) zircon date from upper Adelaidian tuff. Simplified from Powell et al. (1994, fig. 9).

are infirm and are largely based on lithologic similarities and stratigraphic position. It is not at all certain that glacial episodes were of the same age from place to place. Basin margins have been eroded away, largely owing to deformation, uplift, and erosion associated with the subsequent Delamerian orogeny beginning at ca. 540 Ma.

The glacigene beds are well exposed in the Adelaide foldbelt of South Australia, which contains a stratal thickness of more than 10 km (Preiss, 1987). The Adelaidian strata are viewed as first laid down in rifted troughs in several adjoining subbasins and then, as sedimentation continued, over broad regions of subsidence extending beyond rift margins (e.g., Jenkins, 1990; Preiss, 1990; Drexel et al. 1993; Powell, 1995; Walter et al., 1995; Myers et al., 1996). The base of the oldest glacial sequences (Sturtian)

lies above a widespread hiatus and above thousands of meters of Adelaidian strata already lithified at the onset of glaciation. Two incursions of glacially derived sediments are assigned to the Sturtian, separated by units of black shale, siltstone, and dolomite. No undoubted striated basement below tillite has been recognized within the Adelaide sequences; the preserved glacial beds here were laid down within the basin at distance from the margins. At Merinjina Well, however, controversial grooves and striations occur at the base of tillite lying upon volcanic rocks and have been interpreted as of either tectonic or glacial origin (Daily et al., 1973; Preiss, 1987, 1990). If glacial, they may be due to scribing by drifting floes or icebergs.

The lowest Sturtian beds, both marine and nonmarine, grade offshore into ironstone layers suggesting sheltered and perhaps

Figure 54. Glacially sculptured and striated granitic basement below Neoproterozoic Walsh tillite, Kimberley region, Western Australia. Point of knife blade shows direction of ice movement. Centimeter scale. August 1976.

deep water, an environment often associated with a rift (Link and Gostin, 1981; Young and Gostin, 1988, 1989, 1990, 1991). Facies reconstructions indicate that, bordering the southern part of the complex deep Adelaidian trough, glaciers during both Sturtian and Marinoan times lay in rugged mountains to the northeast (Preiss, 1990). In a deep subbasin farther north (Yudnamutana trough), glaciers were located on basement mountains on the southeast and shed debris northwestward during Sturtian time. The chain of elongate Adelaidian basins was therefore intra-cratonic, and strata within them along with basement margins have been deformed subsequently after the ending of Adelaidian sedimentation. They now constitute the Adelaidian foldbelt. Today the eastern edge of the Adelaidian strata is truncated by the Tasman Line and carried away (Fig. 52). Rocks now to the east have been accreted in Phanerozoic time since tectonic truncation.

Across central Australia, glacigene strata lie in remnant basins that include the Georgina, Ngalia, Amadeus, Officer, and Savory basins (Figs. 52, 53). According to Walter et al. (1995), after initiation by broad downwarping to form units of a proposed Centralian superbasin, beginning about 800 m.y. ago, hundreds of meters of marine and fluvial sands were laid down. Walter et al. (1995) suggest that a broad and irregular region of sedimentation was then disrupted tectonically so that a chain of basins originated with Sturtian glaciers sited locally along upfaulted basement highs between them. Marine shales and carbonates were then deposited, followed by renewed glaciation of the Marinoan episodes. Correlation from basin to basin is based on lithostratigraphy and sequence analysis supplemented by recent data from biostratigraphy and some isotopic studies (Walter et al., 1995). The glacial interpretation of diamictites and associated beds across the region as glacigene depends upon faceted and striated stones, dropstones in laminated sequences, and the diversity of erratics. Facies recognized include lodgment till and subglacial and ice marginal deposits, including those with deformed sandstone bodies embedded in poorly bedded diamictite (Lindsay, 1989).

In the Georgina basin the glacigene beds are assigned to the

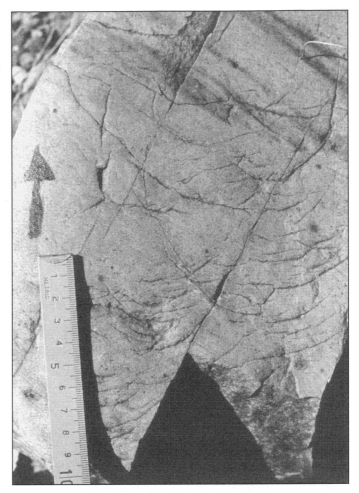

Figure 55. Percussion marks on glacially faceted boulder from Neoproterozoic Walsh tillite, Kimberley region, Western Australia. Ice movement in direction of black arrow. Centimeter scale. August 1976.

Yardiga tillite and Mt. Cornish Formation and are viewed as Sturtian correlatives (Walter 1980). They were deposited in two separate troughs (Keepera and Adam) separated by a basement ridge, and are correlated with the Areyonga Formation of the Amadeus basin to the southwest. Glacigene strata are well preserved in the large Amadeus basin where two lobes of glacial debris, including probable lodgment tills and ice marginal deposits associated with subglacial deformational structures, were laid down along the northeastern margin of the basin. They constitute the Areyonga Formation and its associates (Lindsay, 1989; Korsch and Lindsay, 1989). The material was derived from the region between the Amadeus, Ngalia, and Georgina basins. The eroded southern flank of the Amadeus basin rises against the Musgrave block, a deformed and uplifted region that slowly evolved and reached maximum development during the Cambrian Delamerian orogeny (Preiss, 1987, 1990). The glacial record extends well to the south across the Musgrave block into the Officer basin, where it is preserved as the Chambers Bluff tillite. In the Savory basin to the west, the Boondawari Formation contains glacigene beds, interpreted as derived from glaciers upon highlands to the south, but perhaps also with an input from those upon rising uplands along the trend of the Paterson orogen on the northeast that was originating at this time (Williams, 1992). The glacigene sequences in the central Australian basins are assigned to the Sturtian glaciation, largely on the basis of stratigraphic position and lithologic correlation.

Adjacent to the Kimberley block upper Proterozoic glacigene strata are especially well exposed (Plumb, 1993; Grey and Corkeron, 1998). At least 20 separate localities with glacial pavements are now documented showing ice flow mainly from the present north but also reflecting variations due to subglacial topography. The Landrigan tillite is correlated with Sturtian tillites of the Adelaide foldbelt, the Walsh, Moonlight Valley, and Fargo tillites with the Marinoan tillites; the tillites within the Egan Formation are viewed as somewhat younger. In fact, the Egan may reveal a third glaciation. Each of these glacigene units is capped gradationally by a dolostone layer indicating a sea-level rise following glaciation and perhaps warming of the ambient water.

The ages of the Neoproterozoic glacigene sequences across Australia are not well known largely because of the lack of rocks suitable for radiometric dating in appropriate geological settings and the lack of stable isotopic analyses for use in correlation endeavors. Moreover, paleontological dating is not yet sufficiently refined. The Sturtian within Adelaidian beds of the Flinders Range has an age between 802 ± 10 Ma (a U-Pb zircon date from tuff well below) and 750 ± 150 Ma (Rb-Sr isochron from an overlying shale) (Webb et al., 1983; Fanning et al., 1986; Young and Gostin, 1989). Tentatively, the Sturtian is here placed between ca. 750 Ma and 780 Ma with uncertainty and the Marinoan between ca. 590 Ma and 650 Ma. These uncertain figures give ca. 100 m.y. between the two icy intervals. The Marinoan glacial sequence is somewhat older than a Rb-Sr date of 676 ± 204 Ma from overlying shales (Coats, 1981), and underlies Ediacaran-bearing strata. Although the broad belt of Neoproterozoic basin remnants extends for at least 2,300 km across the continent, the dating as yet does not allow interpretation whether the glaciations were synchronous or diachronous.

The basins display an extensional style, where, following upon broad and irregular downwarping, an early deep rift with uplifted margins was filled with synrift sediments (including evaporites and volcanics), which was then overlapped on the margins by postrift beds according to interpretive models proposed by Jenkins (1990) and Powell et al. (1994). They suggest that a chain of interlinking troughs evolved over an interval of ca. 300 m.y. with five episodes of lithospheric stretching followed by thermal subsidence and widespread flooding and sedimentation between rifting events. This chain lasts a remarkably long time for recurring similar tectonic styles (Fig. 53). Several breaks in the record show that sedimentation was intermittent. The oldest rifting suggests crustal stretching of about three times the thickness of the lithosphere, so that volcanic rocks were emplaced. This occurred at 802 ± 10 Ma, as judged from a U-Pb

zircon date from a tuff within the lower part of the 6 km thickness of the Callanna beds, well below the Sturtian glacial strata (Fanning et al., 1986). At times and places the rifts were filled to above sea level as shown by nonmarine beds and evaporites. At other times, sedimentation surfaces were deep so that ironstone and deep-water shale were laid down.

Only the third and fourth extensional episodes are associated with glacial sedimentation when debris is viewed as derived from nearby sources along faulted and deformed rift margins (Jenkins, 1990; Powell et al., 1994). Between these two tectonic and sedimentation episodes, which resulted in deposition of the Sturtian and Marinoan glacigene units, the widespread Tapley Hill Formation in the Adelaidian fold belt documents a time of regional subsidence extending well beyond rift margins. This interval of subsidence results in the "horns" of the popular "steer's head" analogy (Jenkins, 1990). It was a time between major glaciations. The fourth episode of stretching was associated with the Marinoan glaciation, the episode giving low-latitude paleomagnetic interpretations and periglacial deposits on the marginal shelf (Williams and Tonkin, 1985). The fifth occurred in the Early Cambrian and preceded tectonic events leading into the Delamerian orogeny (Ireland et al., 1998). In summary, debris carried into deep basins to constitute the remarkably thick sequences of Adelaidian strata, laid down during a long interval, was derived from irregular highlands along discontinuous uplifted margins. In Sturtian and Marinoan times, these margins were seemingly high and broad enough to harbor glaciers. In Marinoan time, at places the depositional site was at a low latitude, perhaps as low as 10°, at a time of cool climate (Schmidt and Williams, 1995; Sohl and Christie-Blick, 1995; Sohl, 1997).

Paleomagnetic measurements of Marinoan sequences showing low-latitude glaciation, carried out over many years, are viewed as reliable but questions remain concerning the paleogeography and paleotectonics (McElhinny et al., 1974; McElhinny and Embleton, 1974; McWilliams and McElhinny, 1980; Embleton and Williams, 1986; Idnurm and Giddings, 1988; Schmidt et al., 1991; Williams, 1994; Schmidt and Williams, 1995; Williams et al., 1995; Sohl and Christie-Blick, 1995; Sohl, 1997). The sites giving paleomagnetic information lie within the main Adelaide foldbelt (Elatina Formation) and along its western margin (Whyalla Sandstone) where facies indicate cold climate and cyclicity suggesting tidal control (Williams 1986a, 1988, 1989, 1990; Williams and Sonett, 1985) (Fig. 55). According to Schmidt and Williams (1995, p. 120–121):

> The Marinoan glacial sequence in the Adelaide Geosyncline region includes thick (as much as >1500 m) and extensive (~1000 km) tillite and argillite-dropstone facies containing faceted, polished and striated clasts of various lithologies including dolerite, vesicular basalt, sandstone and dolomite. Faceted clasts show as many as three sets of striae, providing firm evidence of glaciation. Many of the clasts can be traced to sources within the Central Flinders Zone of the Adelaide Geosyncline, indicating the former presence of grounded glaciers in this area. Correlative facies on the Stuart Shelf include permafrost regolith displaying primary sandwedge polygons and other periglacial features such as frost-shat-

tered block-field breccia of as much as 20 m in thickness, frost thrusts, truncated earth mounds, drop involutions, periglacial injections, and frost-heaved blocks (Williams, 1986a). These features of the Marinoan sequence together uniquely define glacial to periglacial environments within the Adelaide Geosyncline region.

The samples giving acceptable paleomagnetic measurements come from both the Elatina Formation within the main trough and two from the periglacial-aeolian Whyalla Sandstone upon the Stuart Shelf, correlated with the Elatina equivalent in the subsurface. The sites are spread over an area measuring ~150^2 km, an area large enough so that precise correlations and structural interpretations are inexact from one site to the next. Those sites within the main Adeladian trough apparently were situated so that they could derive or "cannibalize" clasts from stratigraphically underlying parts of the same Adelaidian sequence, but they are probably not derived from the Stuart Shelf. They must have come from beds within parts of the foldbelt elsewhere, and from outcrops uplifted to erosion at the same time as deposition went on within the main trough. Rifting and pulses of extension, subsidence, and overlapping of faults as sedimentation continued provide plausible mechanisms to explain some aspects of the region's history. Rift margins, uplifted at a time of widespread coolness on Earth, are interpreted as providing sites for glaciers reaching down to the sea at latitudes near 20°.

The western margin of the thick Adelaidian sequence roughly corresponds to the Torrens Hinge Zone (Schmidt and Williams, 1995), the Dundas-Flinders Shear (Jenkins, 1990, fig. 1), and the "G2 gravity corridor" (e.g., O'Driscoll, 1986). No geologist to my knowledge has suggested major strike slip along or beneath the Torrens Hinge Zone, or advocated lateral displacement that might be called upon to separate the ±5° inclinations measured in beds of the main trough from the ±20° inclinations of the Stuart Shelf. However, oblique transtensional opening of the Adelaidian trough from time to time could well be associated with strike slip, as is happening today within the Gulf of California (e.g., Lonsdale, 1989; Dauphin and Simoneit, 1991; Umhoefer et al., 1994). Schmidt and Williams (1995) and Williams et al. (1995) conclude that the Elatina and Whyalla Formations grade laterally from a coastal plain without a nearby rift shoulder. It is plausible that elsewhere glaciers reached the sea to spawn icebergs, some of which may have drifted for great distances toward the equator. If the rift were oriented in a roughly north-south trend, icebergs can be visualized as drifting along it from middle to low latitudes. The distance of travel under this hypothesis would be about the same as the distance of drift today between sites in western Greenland and the Grand Banks off Newfoundland, as diagrammed in Figure 22 (latitude spreads). In addition, perhaps there was a time of quick sea-ice freezing and glaciation followed by rapid warming, as proposed for the Namibian record (Hoffman et al., 1998a,b). Until such concepts are discussed with reference to Australian strata, I hesitate to call upon a scheme to explain near-equatorial ice that involves significant tilting of the Earth's rotational axis as discussed briefly above (Williams, 1993).

In summary, in Australia the Marinoan glaciation or glaciations took place between ca. 590 Ma and 650 Ma during a ca. 60 m.y. interval, and the Sturtian glaciation, some 50 m.y. earlier, between ca. 700 Ma and 750 Ma during a ca. 50 m.y. interval. The dating, however, is not well established. During the Marinoan interval, paleomagnetic data and interpretations show low-latitude depositional sites within part of the Adelaide trough. For Sturtian glaciers reliable paleomagnetic data to indicate latitude of deposition are not yet available. The youngest glacial horizon in the Kimberley region is here interpreted as the result of a young and separate episode, from 545 Ma to 585 Ma, but it may record a long-lasting episode of the Marinoan glaciation.

Western North America

Upper Proterozoic glacigene facies are exposed in several regions extending from Alaska to southern California along the Cordilleran belt of western North America, a stretch of ~3,700 km (Fig. 56). Strong tectonic overprinting along the whole belt, which has to be removed conceptually, however, interferes with satisfactory reconstruction of the geography during Proterozoic times. First, across most of the Cordillera, Mesozoic thrusting of tens or hundreds of kilometers has telescoped stratigraphic sections; and second, later Cenozoic extension has broken and spread them apart again (Armin and Mayer, 1983; Bond et al., 1984, 1985; Levy and Christie-Blick, 1989, 1991; Hansen et al., 1993; Link et al., 1993b). A glacial contribution to the sedimentary record is established, but glaciation along with synsedimentary faulting, downslope sliding, and rapid facies changes confuses interpretations.

In eastern Alaska a glacial episode is documented by facies of the Tindir Group (Ziegler, 1959; Young, 1982; Kaufman et al., 1992; Link et al., 1993b; Klein and Beukes, 1993). In nearby northwestern Canada two glacigene units, assigned to the older Sayunei and Shezal Formations of the Rapitan Group and the younger Ice Brook Formation, are separated by a nonglacial interval, primarily the Twitya Formation (Fig. 57) (Young, 1976b, 1982; Young et al., 1979; Eisbacher, 1981, 1985; Yeo, 1981; Aitken, 1991a,b; Eyles and Young, 1994; Narbonne et al., 1994; Narbonne and Aitken, 1995). Facies distributions indicate mountains adjacent to deep basins with intervening shelf and slope deposits. Repetitions of stacked lithologic types suggest "grand cycles" (Ross 1991; Hansen et al., 1993). Thick deep-water carbonates and iron-bearing units are associated with shallower siliciclastic turbidites. The slope facies include olistostromes and slump blocks intermixed with diamictites and interbedded strata carrying glacial imprints. A capping carbonate, informally known as the "Tepee dolostone" overlies the Ice Brook Formation.

The Neoproterozoic continental margin of North America to the northeast is still preserved and is viewed as having evolved into a passive margin. The matching southwestern margin is interpreted as displaced to central Australia today, where it lies to the west of the Adelaidian foldbelt (Eisbacher, 1985; Young, 1982, 1992a,b, 1995a,b; Christie-Blick et al., 1988; Brookfield,

1993; Powell, 1993, 1995; Powell et al., 1993; Borg and DePaolo, 1994; Eyles and Young, 1994; Idnurm and Giddings, 1995; Rainbird et al., 1996).

The older glacigene sequence of northwestern Canada (Rapitan) is located beneath volcanic rocks of the Mount Harper Complex, which contains zircons dated by U-Pb methods at 751^{+26}_{-18} Ma (Roots and Parrish, 1988; Mustard, 1991; Rainbird et al., 1996). The younger (Ice Brook) lies above this dated volcanic sequence and is interbedded with units containing Ediacaran fossils (first appearing in the underlying Twitya Formation) and grades upward stratigraphically into Cambrian beds (Hofmann, 1987; Hofmann et al., 1990; Narbonne et al., 1994). Both episodes of iciness are placed in Succession C of the Neoproterozoic sequences in northwestern Laurentia (mainly in Canada) (Young, 1982; Link et al., 1993b; Rainbird et al., 1996). The older glaciation is dated somewhere around 750 Ma, and the younger around 600 Ma and ended at ca. 590 Ma. The time interval of ca. 150 m.y. between the two glacial episodes is represented by thick stratigraphic sequences of several different facies that include unconformities and diastems. Paleomagnetic studies show that the Rapitan beds were probably deposited at low latitudes ca. 725 m.y. ago (Morris, 1977; Morris and Aitken, 1982; J. K. Park, 1997).

In the southern Canadian Cordillera, and crossing the border into the United States, thick metasedimentary sequences are reported from several localities within a 350 km strip. Here facies of the Toby, Shedroof, and Huckleberry Conglomerate, including some exposures of thin-bedded argillite with meter-sized lonestones, have been interpreted as laid down in a glacial marine environment (Aalto, 1971). Deposition took place between 762 Ma and 728 Ma along steep rift margins associated with the breakup of western Laurentia (Ross et al., 1995). Most diamictites, however, display no evidence of glacial input (Mustard, 1991). In British Columbia these beds are overlain by strata interpreted as postrift and are succeeded in turn by strata containing a glacigene imprint. Studies of sulfur isotopic changes show sharp variations in sulfates and sulfides that roughly match those of carbon and strontium isotope variations that in turn approximately correspond with glaciations and Ediacaran radiation (Claypool et al., 1980; Holser et al., 1988; Narbonne and Aitken, 1995).

In southeastern Idaho, another 700 km farther south along the Cordilleran chain, the Pocatello Formation crops out in a

Figure 56. Location of Upper Proterozoic glacigene rocks in western North America. Height of column proportional to stratigraphic thickness; see scale. (1) Tinder Group; (2), (3), (4) Rapitan Group; (5) Mount Lloyd George Diamictite; (6) Mount Vreeland Diamictite; (7), (8), (9) Toby, Shedroof, and Huckleberry Conglomerates; (10) Pocatello Formation; (11) Mineral Fork Formation; (12) Sheeprock Group; (13), (14) = Kingston Peak Formation. From J. M. G. Miller, *in* Link et al. (1993b, fig. 36, where details are given). AB = Alberta; AK = Alaska; AZ = Arizona; BC = British Columbia; CA = California; CO = Colorado; ID = Idaho; MEX = Mexico; MT = Montana; NM = New Mexico; NV = Nevada; OR = Oregon; SK = Saskatchewan; UT = Utah; WA = Washington; WY = Wyoming; YT = Yukon Territory.

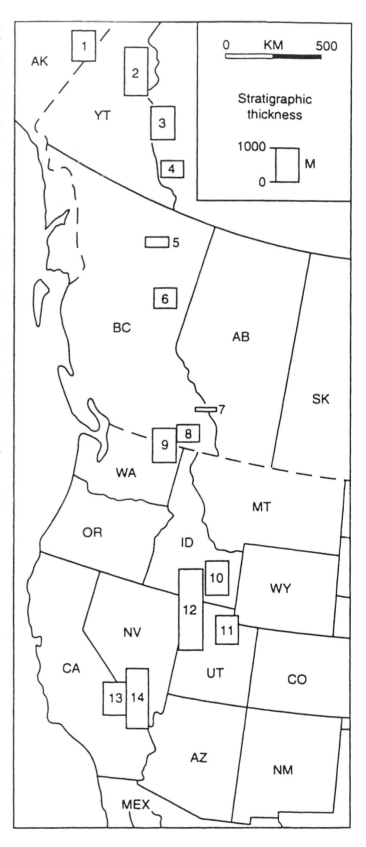

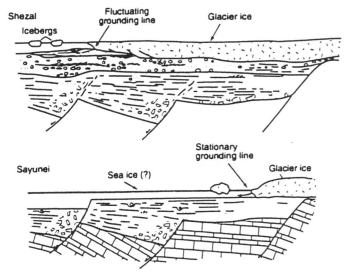

Figure 57. Depositional environments of the central facies, Neoproterozoic Rapitan Group, western Canada. From Eisbacher (1985, fig. 8).

100 km strip near the Utah border (Link, 1983; Link et al., 1994). The formation, nearly 2,000 m thick, contains the Scout Mountain Member interpreted as glacial marine on the basis of massive diamictites containing rare glacially striated clasts. The interbedded Bannock Volcanic Member, as yet undated, suggests a rift setting (Harper and Link, 1985). The stratal sequence, including a capping carbonate, points toward the filling of an irregular rift system accompanied by later subsidence. Overlying the glacigene units are about 6 km of uppermost Proterozoic beds, largely quartzites. Strata here as elsewhere along the Cordilleran trend are allochthonous, severely folded and faulted, and have been carried eastward by Mesozoic thrusting for tens of kilometers to now lie upon the margin of the continental craton. They have since been tectonically overprinted by later deformation, largely extensional.

In Utah, the Mineral Fork Formation and its approximate correlatives (Dutch Peak, Otts Canyon, Horse Canyon Formations, and "Formation of Perry Canyon") occur within and near the Wasatch Mountains, the Sheeprock Mountains, and islands of Great Salt Lake. Two intervals of glaciation are inferred. Evidence for glaciation includes coarse, thick diamictite layers mainly interpreted as of glacial-marine origin (Crittenden et al., 1983; Christie-Blick 1982a, 1983; Link et al., 1993b, 1994). At places laminated shale contains dropstones as large as 2.5 m along with till clots. Within the central Wasatch Range, the Mineral Fork Formation lies upon a glacially striated pavement with roches moutonnées. Within this region as well, two glacial canyons have been identified (Ojakangas and Matsch, 1980, 1982; Christie-Blick 1982b; Link et al., 1993b, 1994). Paleogeographic reconstruction of the piecemeal record suggests an irregular cratonic margin with local terrestrial deposits still preserved, opening westward into marine waters (Christie-Blick et al., 1988; Levy and Christie-Blick, 1989, 1991). Overlying the

glacigene beds is a latest Proterozoic section ~4,000 m thick, largely quartzite, that includes units that are satisfactorily correlated with those in southern Idaho.

Still farther south ~600 km along the Cordilleran belt in the Death Valley region of southeastern California, the Kingston Peak Formation (as much as 3,000 m thick) is well exposed in the desert environment (Johnson, 1957; Miller, 1985, 1987; Miller et al., 1988; Wright and Prave, 1993; Link et al., 1993a,b). Here again diamictite and other facies, including those with rare glacially striated stones, indicate glaciers upslope somewhere in a tectonically active environment (Walker et al., 1986). Conglomerates and breccias, some of which contain huge allochthonous blocks and olistostromes, grade laterally into deep-water deposits, now argillites with some dropstones. Islands lay on the present west of the irregular cratonic margin, and some sediment source areas lay to the present south. Syntectonic downslope sliding took place from rugged basin margins that probably harbored glaciers locally from time to time. The stratigraphy of the Kingston Peak Formation is complicated; the middle and upper part includes scarp-derived facies that herald the inception of rifting and lie unconformably above the lower part containing undated altered pillow basalts (Miller, 1985; Walker et al., 1986; Heaman and Grotzinger, 1992).

The youngest Neoproterozoic diamictites in the Death Valley region, interpreted as lodgment tillite, locally interfinger with the lower part of the overlying Noonday Dolomite, which at places overlaps the Kingston Peak Formation and inferred buried faults (Miller, 1985). The Noonday is the lowest unit of a nearly 2,000 m thickness of latest Proterozoic postrift strata lying below the base of the Cambrian (Levy and Christie-Blick, 1989, 1991; Elston et al., 1993; Wright and Prave, 1993; Corsetti and Kaufman, 1994). These units have yielded stromatolites and complex metazoans, but the fossils are not yet useful in wide correlation. Based upon regional relations, deposition of the Kingston Peak Formation and concurrent and intermittent episodes of basin-margin glaciation are judged to have an age between 780 Ma and 700 Ma. The glacigene beds at the Kingston Peak–Noonday contact zone may record a second and younger glaciation (Miller, 1985; Wright and Prave, 1993; Link et al., 1993b); Heaman and Grotzinger (1992) suggest a date as young as 600 Ma. Efforts to find reliable paleomagnetic pole positions have so far been unsuccessful (Elston et al., 1993). Correlations and dating within this complex region are still quite uncertain, however, because the region has experienced separate episodes of strong Mesozoic thrusting and marked Cenozoic extension, all superposed upon the record of complicated Proterozoic tectonic events that disrupted the continental margin (Burchfiel et al., 1992; Horodyski, 1993). No exposures of glacigene rocks along the Cordillera southeast of the Death Valley region have been confirmed although diamictites along with some dropstones are reported in the Florida Mountains of southern New Mexico (Corbitt and Woodward, 1973).

Along the Neoproterozoic rifted and irregular and rugged western margin of North America from easternmost Alaska to southeastern California, data from sequence stratigraphy, subsidence analysis, variations in geochemical parameters, and regional stratal characteristics provide a tentative picture (Kaufman et al., 1992; Wickham and Peters, 1993; Narbonne et al., 1994; Knoll and Walter, 1995). Local glaciers surmounted the irregular continental margin during two rifting episodes, the first between 780 Ma and 700 Ma and the second at ca. 600 Ma, but it is not known how extensive glaciers were during their maximum waxings. The uncertainty of correlations and timings, however, permits marked diachroneity of both tectonic events and glaciation during this span of ca. 180 m.y. along the ~3,700-km-long belt. During the first episode the northwestern Canadian part of the belt probably lay at low latitudes (J. K. Park, 1997).

Eastern North America and northern Europe, including Greenland

Upper Proterozoic stratal sequences displaying a glacial contribution lie on either side of the current North Atlantic Ocean (Fig. 58). They are interpreted as laid down both along the eastern margin of the Laurentian craton and along the western margin of the Baltic craton during stages when the united craton first began to rift apart to form the Iapetus Ocean in early Paleozoic time.

In the central Appalachian Mountains of southwestern Virginia, glacigene beds are assigned to the Konnarock Formation (formerly the upper Mount Rogers Formation) (Rankin, 1975, 1993; Rankin et al., 1993; Miller, 1994). These metamorphic rocks (lower greenschist facies) lie within thrust sheets that were tectonically transported northwestward during the Paleozoic Appalachian orogeny, tectonic events that closed the Iapetus Ocean. The glacial interpretation depends on the occurrence of massive diamictite layers; dropstones, till clasts, and pellets embedded within laminated sequences; and laminites with varve-like appearances (Miller 1994). The age of rhyolites in the Mount Rogers Formation, immediately underlying the Konnarock Formation, is 759 ± 2 Ma based on U-Pb dating (Aleinikoff et al., 1991). Trace and body fossils, along with dated basalt flows, in the overlying Chilhowee Group of the Neoproterozoic place the age of the glaciogenic rocks between 760 Ma and 570 Ma. Reconstruction of the paleotectonic setting suggests deposition of the Konnarock Formation within a small basin with significant relief on a continent, during or shortly after rifting (Miller, 1994). The glaciers were probably alpine or valley glaciers, and extended into a nearby lake. Fresh water is inferred because associated strata, including red beds, are viewed as nonmarine. Some of the diamictites may record material dropped from floating ice near shore, others may be lodgment tillite, and still others may have slid into place on a subaqueous slope.

In southeastern Newfoundland the glacially influenced Gaskiers Formation (Conception Group) consists primarily of diamictites interbedded conformably within turbidites (Eyles and Eyles, 1989; Eyles, 1990; Myrow, 1995). The formation includes as much as 300 m of massive and crudely stratified diamictite along with broken diamictite layers carrying folded and disrupted sediment rafts. The diamictite layers are interbedded within thin turbidite and mudstone sequences indicating that they were deposited in a subaqueous environment as debris flows upon a slope or at the base of a slope. Their lateral continuity suggests that the debris came from a number of unstable slump sources that lay at an unknown distance from glacier fronts. Similar units, tentatively correlated with the Gaskiers Formation, extend about 125 km north and south and 275 km east and west. Glaciers contributed to the debris as shown by rare glacially striated and shaped stones, and dropstones are reported. Because much of the debris is volcanic, it is inferred that deposition took place along the base of a volcanic chain, but only a small proportion of the debris is pictured as derived from glaciers sited upon the volcanoes themselves. The Gaskiers Formation is younger than $606^{+3.7}_{-2.9}$ Ma, older than 565 ± 3 Ma from U-Pb zircon ages (Myrow, 1995), and conformably underlies ~2,000 m of strata below Ediacaran fossils (Anderson, 1972). The formation is encompassed within the Avalonian-Cadomian orogenic belt formed when the Iapetus Ocean closed, and is interpreted as allochthonous to North America.

Along the coasts of Greenland ancient glacigene strata are well exposed (Fig. 58). In east Greenland, two diamictite-bearing sequences (Ulvesø and Storeelv Formations) occur within the 17-km-thick section of Late Proterozoic beds, separated and overlain by marine mudstones and sandstones (Hambrey and

Figure 58. Neoproterozoic basins containing glacigene record along the early Iapetus rift passing through Laurentia. After Nystuen (1985, fig. 2) and Eyles (1993, fig. 11.1).

Spencer, 1987; Moncrieff and Hambrey, 1988, 1990; Hambrey and Moncrieff, 1985; Manby and Hambrey, 1989; Moncrieff, 1989). The basal tillite facies is interpreted as a subglacial and lodgment deposit. Pebble fabrics display strong alignment and at least two glacially striated boulder pavements occur that were probably formed by a glacier grounded below sea level. Higher in the section dropstones and till pellets are identified (with some uncertainty) along with additional boulder pavements lacking striated tops with preferred orientations. The excellent exposures allow positioning of glacier grounding lines and transitions from basal lodgment tillite to waterlain tillite seaward. In addition, debris-flow deposits are recognized with erosional bases, rip-up clasts, and flow-banding within a homogeneous matrix. Glacial-marine facies include those that are proximal to the ice front, and those more distal where the glacial input is only rarely documented by clusters of clasts presumably dropped by individual icebergs as they tipped or melted. Other facies recognized are rhythmites (including turbidites), and aeolian, permafrost, and evaporite deposits. Widespread glaciers with broad floating shelves contributed large amounts of sediment. Basement highlands lay on the present west and southwest, and harbored glaciers extending into the opening Iapetus seaway.

In northern Greenland the Franklinian mobile belt, with an orientation at right angles to the opening Iapetus Ocean, the stratigraphic and tectonic arrangements suggest deposition within an aulacogen. A thick section of glacigene strata, assigned to the Moræ_nesø Formation, was derived from rift shoulders that display block faulting (Surlyk, 1991; Eyles, 1993). Diabasic dike swarms, only tentatively dated, suggest that rifting was underway by ca. 750 Ma (Trettin, 1989). Glacial paleovalleys containing diamictites carrying striated and faceted clasts are preserved in Peary Land, north Greenland (Collinson et al., 1989).

In Spitzbergen a Proterozoic glacial record has long been recognized (Garwood and Gregory, 1898; Wilson and Harland, 1964). Here, and in other parts of the Svalbard Archipelago, glacigene rocks are well exposed and include diamictites with faceted and striated stones that are both basinal and extrabasinal in provenance. Within thin-bedded units such stones also occur along with rhythmites carrying diamictite pellets, and periglacial structures such as wedge fillings are also reported (Chumakov, 1968; Fairchild et al., 1989; Harland et al., 1993; Fairchild and Hambrey, 1995). In Spitzbergen, terrestrial facies include beds laid down as outwash and lacustrine deposits as well as turbidites, some of which reveal bottom-current reworking. At several places cap carbonates loaded with dolomitic detritus eroded from substrate crop out. These occurrences are important in interpreting the significance of the carbonate-tillite association (Fairchild et al., 1989; Fairchild, 1993).

In reconstructing the far–North Atlantic region by closing the Iapetus sea and removing Caledonian deformation, both the north and east Greenland regions are placed close to Spitzbergen and other parts of the Svalbard Archipelago. Glacigene sediments in all three regions were probably laid down in closely associated basins (Hambrey et al., 1981; Anderton, 1982; Nystuen, 1983,

1985; Surlyk, 1991; Harland et al., 1993; Fairchild and Hambrey, 1995). Two marked glacial episodes are documented that suggest advancing and retreating of glaciers accompanied by the rise and fall of sea level. The icy episodes, judging from stratal thicknesses, constituted no more than 15% of Vendian time according to an estimate by Harland et al. (1993).

In northwestern Europe, Neoproterozoic sites with glacigene aspects have been under study intermittently since Thompson in 1871 and 1877 noted glacial deposits on Islay, Scotland, although at that time their Proterozoic age was not yet established (Harland et al., 1993). In 1891 Reusch described a "moraine" with an underlying striated pavement along the shore of Varangerfjord, northern Norway—the notable "Bigganjarga tillite" or "Reusch's moraine" (Figs. 59, 60). Studies for well over a century now show that the deposits were laid down in a marginal marine setting within a glacially influenced fan delta system, largely by debris flows but accompanied by ice-rafting in distal beds (Holtedahl, 1918; Føyn, 1937, 1985; Crowell, 1964; Spjeldnæs, 1964; Reading and Walker, 1966; Bjørlykke, 1967; Laird, 1972; Edwards and Føyn, 1981; Edwards, 1984, 1997; Gayer and Rice, 1989; Siedlecka and Roberts, 1992; Drinkwater et al., 1996; Jensen and Wulff-Pedersen, 1996, 1997). "Reusch's moraine" is now interpreted as a diamictite mass emplaced as a viscous debris flow, in part of probable glacial origin, near the base of the Smalfjord Formation where this formation laps upon the Baltoscandian craton. As the debris flow moved into position, it scribed firm but unindurated sand layers beneath it (now quartzites). A few grooves in this pavement retain pebbles at their downflow ends and display a scribing trail corresponding to the pebble shape. Rainout diamictites and debris clusters are identified in associated facies, and faceted and glacially striated clasts noted (Edwards and Føyn, 1981). The overlying Nyborg Formation includes turbidites that document marine transgression, and the still higher Mortensnes Formation documents an additional probable advance or recurrence of glaciation on nearby land. Two glacial episodes are therefore suggested. Paleomagnetic results are as yet unsatisfactory but tentatively suggest deposition in mid- to high latitudes (Vidal and Bylund, 1981). Isotopic dating is insecure although dates of 654 ± 23 Ma (recalculated Rb-Sr) are reported from the Nyborg Formation (Pringle, 1973; Hambrey and Harland, 1981). Dikes dated at ca. 640 Ma (potassium-argon methods by Beckinsale et al., 1976) and block faulting along the margin of a probable aulacogen are interpreted as heralding crustal extension associated with Iapetus opening (Eyles, 1993).

In central Scandinavia, strata with a glacial imprint, including the Moelv tillite, are preserved in at least five irregular downfaulted basins (Sparagmite basins). These basins originally lay at or near the cratonic margin but have now been carried eastward by Caledonian thrusting (Bjørlykke, 1966, 1985; Nystuen and Sæther, 1979; Bjørlykke and Nystuen, 1981; Sæther and Nystuen, 1981; Thelander, 1981; Vidal and Bylund, 1981; Nystuen, 1983; Kumpulainen and Nystuen, 1985; Eyles, 1993; Vidal and Moczydłowska, 1995). Debris flows, some carrying

Figure 59. Neoproterozoic Bigganjarga tillite (Reusch's moraine), Varangerfjord, northern Norway. August 1960.

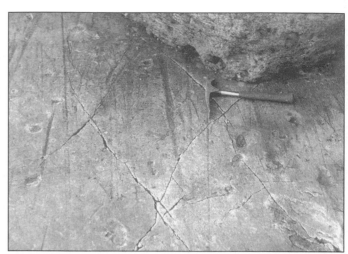

Figure 60. Grooved and striated bedding-plane surface beneath Bigganjarga tillite; same locality as in Figure 59.

boulders more than 1 m in size, are associated with faults along basin margins and distally are interbedded with turbidites. Some clasts exhibit glacial facets and striations and some were emplaced as dropstones into thin-bedded sequences (Nystuen and Sæther, 1979). Facies studies indicate deep basins with rugged margins, here and there with glaciers, but tentative reconstructions do not require broad ice sheets.

Vendian beds in central and eastern Europe are reported to show glacial influence at localities extending from the Kola Peninsula on the north to the southern Ural Mountains (Hambrey and Harland, 1981; Linnemann, 1995). Drill holes into the subsurface Orsha basin, in Belarus southeast of Minsk, reveal facies documenting glaciation associated with lacustrine and fluvial deposits, laid down within shallow graben or rifts (Bessonova and Chumakov, 1968). Ice was present from the Baltic region across the East European Platform, but it is not known whether it constituted a continuous ice sheet or local centers and caps. Dating of the record is poor.

In Normandy, especially in coastal exposures near Granville, conglomerates, sandstones, and diamictites with deformed interbeds have not been shown to be glacial in origin (Winterer, 1964; Graindor, 1965; Doré, 1981; Doré et al., 1985). Striations on stones are the result of rotation and scribing within an abrasive matrix during sliding and tectonic deformation.

In Scotland, especially in the Inner Hebrides Islands, Proterozoic glacial and associated facies are well exposed (Kilburn et al., 1965; Spencer, 1971; Eyles and Eyles, 1983; N. Eyles and Clark, 1985; C. H. Eyles, 1988; N. Eyles, 1993). Here the Port Askaig Formation consists of ~750 m of diamictite interbedded with laminated siltstone, sandstone, conglomerate, and carbonate rocks probably laid down upon a shallow marine shelf with glaciers nearby. The beds include lenticular and deformed diamictites, olistostromes with dolomite blocks, and massive mudstone all laid down with rapid facies changes, and are associated with turbidites. Dalradian ice-borne boulders are also reported from

Banffshire, northeastern Scotland (Sutton and Watson, 1954). The belt of these Dalradian beds extends from mainland Scotland into northwestern Ireland (Kilburn et al., 1965; Howarth et al., 1966; Howarth, 1971; Max, 1981; Treagus, 1981; Evans and Tanner, 1996). Dropstones within thin-bedded units include rock types similar to basement rocks in nearby areas (Evans et al., 1998). These and the general characteristics of the facies indicate glaciation and cold climate in the region, although no glacially faceted and striated stones, or pavements, have been reported from these deformed and metamorphosed beds.

Asia

In China, from near the South China Sea and westward for ~4,000 km across Sinkiang Province (Xinjiang), three Neoproterozoic glacial sequences are placed within the Sinian system, but their dating is only approximate: (1) Chang'an or Gucheng (ca. 800–760 Ma), (2) Nantuo (ca. 720–680 Ma), and (3) Luoquan (ca. 620–600 Ma) (Liu Hung-yun et al., 1973; Shih-fan Liao et al., 1976; Chen Jinbiao et al., 1981; Mu Yongji, 1981; Shih-fan Liao, 1981; Wang Yuelin et al., 1981; Lu Songnian et al., 1985; Zhang and Zhang, 1985; Guan Baode et al., 1986; Lu Songnian and Qu Lesheng, 1987; Zheng Zhaochang et al., 1994; Li et al., 1995, 1996; Rui and Piper, 1997). The stratigraphy and tectonics across the huge region are complex and correlations between the separated regions, which may originally have been farther apart or closer together as tectonic blocks have moved about, are quite tentative. The glacigene beds are associated with three major tectonic blocks: (1) Tarim, in the far west, (2) North China, and (3) South China, including the Yangtze and Cathaysia blocks (Lin et al, 1985; Şengör et al., 1988; Li et al., 1995, 1996; Rui and Piper, 1997). In addition, other blocks are proposed in central Asia that include the Kazakhstan, Tuva-Mongolia, and central Mongolia blocks (Ilyin, 1990). The original position of these blocks and their subsequent tectonic and sedimentation histories remain under investigation.

At places on each of these blocks there is evidence of glaciation such as diamictites carrying striated, and polished, stones embedded in massive units, as well as thin-bedded intercalations with dropstones (personal observations at limited localities on North and South blocks, October 1981). From north to south in central eastern China, some examples are: (1) On the North China block terrestrial deposits, including red beds, of the glacigene Luoquan sequence in Shanxi Province, interpreted as lodgment tillite, pass laterally into subaqueous deposits over a distance of ~200 km (Fig. 61) (Guan Baode et al., 1986). Here ice-flow and current-flow directions are variable and include glacial striations and furrows, scribed into firm sand that is now indurated. (2) Upon the South China block to the southeast, in the vicinity of the Yangtze Gorges, facies of the Nantuo Formation are interpreted as showing transition from terrestrial tillite into marine deposits over a distance of ~50 km (Lu Songnian and Qu Lesheng, 1987). These are placed high in the Sinian system. Some 95 km to the southwest of the Yangtze Gorges localities, facies of the Nantuo Formation include periglacial features such as probable ice wedges (Qi Rui Zhang, 1994). Stratigraphic units of both the Nantuo and Chang'an Formations are placed on the South China block. In Yunan Province more than 1,200 km farther to the southwest, trenches at Wangjiawan near Kunming display convincing glacial dropstones within laminated shale. (3) Along the northeastern margin of the Tarim block far to the west in Sinkiang (Xinjiang), three formations with a glacigene contribution are shown in stratigraphic columns by Li et al. (1996). These investigators correlate the uppermost glacial unit on the Tarim block with the Luoquan Formation of the North China block and with the Nantuo Formation of the South China block, respectively. Probable glacigene beds are reported on several of the blocks in central Asia, including Mongolia (Ilyin, 1990; Brasier et al., 1996; Lindsay et al., 1996) and perhaps also on the deformed northern margin of Indostan beneath the complex Himalayan belt (Brasier and Singh, 1987).

Paleomagnetic studies, coupled to geologic reconstructions, suggest that the three major tectonic blocks (North China, South China, and Tarim) record Neoproterozoic rifting that broke up the supercontinent of Rodinia (Powell et al., 1994; Li et al., 1996). Comparisons between stratigraphic sections, accompanied by paleomagnetic data, tentatively place the South China block and Australia adjacent to each other at ca. 700 Ma and have since been separated beginning with the fragmentation of the supercontinent of Rodinia. China and Australia are located near northwestern Laurentia at these times, long before several later tectonic events, including the Mesozoic opening of the Pacific Ocean. The Tarim block at ca. 700 Ma is shown at ~60° N lat, separate from the others that are clustered near 30° N lat (Li et al., 1996).

The ages of these glacial sequences in China depend primarily on stratigraphic position, a few fossils, and some K-Ar and Rb-Sr isotopic dates, mainly from glauconite (Lu Songnian et al., 1985; Guan Baode et al., 1986; Li et al., 1995).

Figure 61. Dropstone in thin-bedded sequence, Neoproterozoic Sinian series, Luoquan Formation, Shangzhanquan locality, Lonan County, Xi'an District, Shanxi Province, China. Coin is ~2 cm in diameter. October 1981.

Ediacaran-like fossils are associated with Luoquan glacigene beds lying a few meters stratigraphically below fossiliferous Cambrian beds and occur above the Nantuo glacial beds (younger than ca. 680 Ma), and at Wangjiawan (near Kunming, Yunan Province). The youngest glacigene sequence is therefore accompanied by Ediacaran fossils and the older sequences are below them. In south China fossil algae and animal embryos are preserved in phosphorite layers just below the Ediacaran radiation (Xiao et al., 1998). The age of oldest glacial sequences, such as those of the Chang'an (Gucheng) (ca. 700–760 Ma) is even less well established. Interregional correlation of Neoproterozoic sequences employing carbon and sulfur isotopic excursions across Asia are underway (e.g., Kaufman and Knoll, 1995).

South America

Neoproterozoic glacial deposits are identified in at least two regions far apart in central Brazil, at localities on the west near the Bolivian border and those in northern Minas Gerais state, ~1,600 km to the east. In the western or Mato Grosso region, northwest of Cuiabá, clasts with glacial facets, polish, and striations, and some "bullet-boulders" occur in diamictites and in turbidites as dropstones. These beds are assigned to the Jangada Group and the Puga Formations (Rocha-Campos and Hasui, 1981). Here terrestrial tillite, deposited on the margin of the Amazonian craton that lies to the northwest, grades laterally over ~200 km into deep-water turbidites with glacial dropstones (Alvarenga and Trompette, 1992). Regional correlation of these Cuiabá beds into those carrying an Ediacaran-like fauna, and a younger crosscutting granite giving a whole rock K-Ar and Rb-Sr date of ca. 500 Ma (Almeida and Mantovani, 1975) indicate a youngest Neoproterozoic age for this glacial episode. The glaciation here was located on the divergent margin of the Amazon craton with the Paraguay basin, and perhaps related to the initial rift opening of the Iapetus Ocean (Trompette, 1994; Grunow et al., 1996). Strata along this margin were later deformed and metamorphosed within the Paraguai belt during the Brazilian orogeny ca. 600 Ma (Pimentel and Fuck, 1992; Trompette, 1994).

About 1,300 km east of the Cuiabá region, a record of an older glaciation is well exposed in the Jequitai area within the Macaúbas Group of central northern Minas Gerais state and extending across the border into the state of Bahia (Isotta et al., 1969; Pflug and Schöll, 1975; Hettich, 1977; Rocha-Campos and Hasui, 1981). At Jequitai glacigenic diamictite and associated facies containing glacially shaped, faceted, striated, and polished clasts lie upon a grooved and striated pavement scribed and furrowed when a glacier moved across firm but unindurated sand (Figs. 62–64). The Macaúbas Group underlies the thick Bambui Group at the top of the Neoproterozoic section in this region. A metarhyolite within the deformed Macaúbas sequence gives a U-Pb zircon date of 794 ± 10 Ma. The glaciation is therefore probably as old as 800 Ma but with large uncertainty and is placed at ca. 900 Ma by Trompette (1994). The beds were deposited at sites flanking the São Francisco craton, perhaps associated with rifting, but are now deformed within the Uruçu and Brazilia orogenic belts, which in turn are parts of the extensive Pan-African system. Rock slices of oceanic affinity within the Uruçu belt suggest that an ocean basin of unknown width closed between the Amazon and São Francisco cratons, bringing the Cuibá and Jequitai regions closer together. In summary, it seems likely that two Neoproterozoic glaciations are recorded in Brazil: one occurring on the west just before the transition into the Cambrian and another far to the east at ca. 900 Ma.

Trompette (1994) relates the glaciations to multiple and diachronous rifting, or to the formation of foreland basins, in three main stages: (1) An old episode, poorly dated and perhaps as old as 1000 Ma, recorded within complicated foldbelts in central Brazil. These belts, reactivated at times, lie between the Amazonian craton and the São Francisco craton; the latter is depicted as the South American part of the African Congo craton. Glaciation is interpreted as both preceding ancient rifting and accompanying it. (2) Between 850 Ma and 900 Ma, where the

Figure 62. Furrowed and grooved glacial pavement. Neoproterozoic Macaúbas Group, Jequitai area, northern Minas Gerais state, Brazil. Scale shown by distant hammer leaning against overlying tillite, illustrated in Figure 63. July 1975.

Figure 63. Neoproterozoic Jequitai tillite, locality 1 km north of locality in Figure 62. July 1975.

dating is plausibly in accord with correlation to African events. This episode is preserved in deposition of the Macaúbas Group in Minas Gerais. (3) Between 550 and 650 Ma, when the Jangada Group of Mato Grosso was laid down.

Antarctica

Probable glacigene metadiamictites occur in the Nimrod Glacier area of the central Transantarctic Mountains (Stump et al., 1988). These rocks have been metamorphosed and deformed so that original sedimentologic details are obscure. Note that Stump et al. (1988) consider the glaciation as "possible" rather than "probable." The metagraywackes and phyllites include thin-bedded units with lonestones showing depressed laminations, but no faceted or striated stones have been recognized. These beds are known only to be Late Proterozoic in age, based on their stratigraphic position.

Summary and discussion: Late Proterozoic glaciations

Several Neoproterozoic glacial episodes are well established although confidence limits on their dating are wide. Two intervals are especially well documented (Kennedy et al., 1998). A younger interval, documented in northern Europe and vicinity (Varangian) and in Australia (Marinoan) are placed between 590 Ma and 640 Ma. These labels are commonly extended over the Earth. An older time of glaciations is placed between 700 Ma and 750 Ma where it is especially well documented in Australia (Sturtian). The Sturtian label is also extended over the Earth, including Africa and North America. There may be a still younger episode between ca. 545 Ma and 585 Ma (perhaps as young as 520 Ma) where glacigene strata overlie beds carrying the Ediacaran fauna in northern Africa, Brazil, China, northwesternmost Canada, and perhaps in northwestern Australia. The oldest Neoproterozoic glacial interval, judged to be ca. 900 Ma or older, as tentatively dated, is reported in central Africa and Brazil (Trompette, 1994). Investigations of stratal sections for most of these episodes show that there is little question of a glacigene origin for some of the beds, but questions remain concerning the age and tectonic, paleogeographic, and paleolatitude settings.

Because time limits of the main Neoproterozoic ice ages as now known are so broad—ca. 50 m.y.—tectonic arrangements may well have changed during the interval. These may include the rising and wearing away of mountain ranges, or the opening and closing of seaways between islands and continents. Glacial centers may have moved about leaving diachronous deposits as they did during Phanerozoic time. Only within broad limits can the glacigene deposits be used for correlation (Crowell and Frakes, 1970; Crawford and Daily, 1971; Kröner, 1977; Deynoux et al., 1978; Chumakov, 1981, 1985; Ojakangas, 1988). Modern geochemical studies suggest, however, that marked isotopic excursions reveal widespread and short events related to glaciation (Derry et al., 1992; Knoll and Walter, 1992; Kaufman et al., 1997; Hoffman et al., 1998a,b; Kennedy, 1996; Kennedy et al.,

Figure 64. Glacially faceted and striated cobble. Same locality as in Figure 63. Knife is 8.5 cm long. July 1975.

1998). Investigations of far-separated stratigraphic sections are now needed similar to those in Namibia where isotopic dating is available if appropriate rocks and methods can be found (Saylor et al., 1995, 1998; Hoffman et al., 1998a,b). Until such studies have been undertaken and demonstrate well dated events bearing on climate change, it is prudent to label past ice ages only by approximate dating and to name them only in accord with the local region of their exposure. I view it as premature to attach names such as Sturtian, Marinoan, and Varangian to ice ages in regions far from their type localities and to imply a worldwide correlation. It is likely that the glaciations were diachronous. Their widespread and near-synchronous occurrence may be proven soon, however, because several critical studies are underway. These primarily involve the measurement and interpretations of isotopic excursions.

Much focus is now placed upon making better reconstructions of continental arrangements for Neoproterozoic time and these are expected to improve through the years ahead (Lindsay et al., 1987; Hoffman, 1988, 1989a,b, 1991; Scotese, 1990; Van der Voo and Meert, 1991; Dalziel, 1991, 1994, 1997; Moores,

1991; Kirschvink, 1992; Powell, 1993; Powell et al., 1993, 1994; Meert and Van der Voo, 1994; Dalziel and McMenamin, 1995; Torsvik et al., 1995, 1996; Li et al., 1996; Myers et al., 1996; Kirschvink et al., 1997). On most of these maps, glacial symbols are emplaced only approximately with no thorough treatment of stratal facies gradations and local tectonic and geographic settings. Long-distance transport of glacial debris by icebergs is usually not appraised. Moreover, the premise of glaciation occurring in high latitudes is often basic to the reconstruction at hand, and it is not always clear to what extent the reconstruction also depends on this premise. Most reconstructions are rightly based on paleomagnetic data, but these, even if quite reliable, give no information on longitude. And yet, in reconstructing the past distribution of land and sea, there are never enough data; so it is essential to use all information available, but to use it judiciously. Such difficulties underline the requirement in historical reconstructions not only to use wisely all information available, but also to show the limits of applicability of each component. These components include the fitting of shapes of cratonic blocks, paleomagnetic measurements, dates, paleogeographic reconstructions from studies of stratal facies, trends in pre-breakup structures, and others.

During Neoproterozoic time, the supercontinent of Gondwana had not yet assembled. It was formed by the gathering together of cratonic blocks at the end of the Neoproterozoic, and the assembly was completed by the time of the Cambrian to Ordovician transition. Details of Gondwana's evolution are held within deformed older rocks of the Pan-African orogen that may reveal when previous oceans existed and were then closed (Li and Powell, 1993; Trompette, 1994; Veevers and Powell, 1994; Torsvik et al., 1996; Kirschvink et al., 1997). Reconstructions proposed for Neoproterozoic time, although still quite sketchy and preliminary, suggest that a previous supercontinent made up mainly of blocks later incorporated into Gondwana existed between 545 Ma and 580 Ma (Li et al., 1996; Dalziel, 1997; Unrug, 1997). This supercontinent, named *Pannotia* (Powell, 1995; Dalziel, 1997; Unrug, 1997; but see also Dalziel and McMenamin, 1995; Young, 1995a,b), came into existence after Gondwana had partly assembled. Laurentia is viewed as joined to part of Gondwana, to rocks now in southern South America (Hoffman, 1991, 1992; Dalziel, 1994; Bond et al., 1984; Sadowski and Bettencourt, 1996). This joining is in accordance with the "SWEAT" hypothesis (Dalziel, 1991, Moores, 1991).

The Late Proterozoic supercontinent previous to Pannotia, named *Rodinia*, predominated the Earth's geography from ca. 750 Ma to 1000 or 1100 Ma (McMenamin and McMenamin, 1990; Unrug, 1996, 1997). Wide oceans and changing shapes and positionings with continental islands within them are viewed as characterizing this long interval of as much as 170 m.y. between the breakup of Rodinia and the formation of Pannotia. It is noteworthy, as pointed out by several investigators, that the Sturtian glaciation in Australia occurs during the breakup of Rodinia and the Marinoan (Australia) and Varangian (northern Europe) at some time before Gondwana finally assembled. The likely youngest Neoproterozoic glaciation corresponds closely in time with the existence and breakup of Pannotia, a supercontinent that may have been first relatively high standing with respect to sea level and then was fragmented so that sea level encroached upon the margins of the pieces. The association of these major tectonic events and times of widespread iciness on Earth is viewed as related, but whether through the influences of continental positioning or events causing changes in the tectonic flux of greenhouse gases from the Earth's interior or other factors is not clear. The end of the Proterozoic was a time of dramatic changes in biological activity and their geochemical effects and in the style and pace of tectonic events (Schopf, 1992, 1994; Schopf, 1982; Knoll and Walter, 1995; Logan et al., 1995; Kaufman et al., 1997; Kennedy et al., 1998). It may even have been a time when the whole crust slipped rapidly with respect to the deeper Earth, a time of "true polar wander" (Kirschvink et al., 1997).

Basic to all modern reconstructions is the premise that tectonic processes similar to those associated with Phanerozoic plate tectonics also operated in the Proterozoic. Cratonic blocks moved about on a mobile crust and oceans opened and closed. Marked changes in the arrangement of cratonic blocks may have brought about marked changes in climate as well. For example, Kirschvink (1992) suggests that equatorial clustering of most or all continental blocks may have been largely responsible for episodes of low-latitude glaciation during the Neoproterozoic. This is viewed as enhancing deep tropical weathering and carbon dioxide drawdown. Moreover, without adequate continental blocks in polar regions, glaciers cannot grow there. In addition, Kimura et al. (1997) propose that there was oceanic overturning in latest Neoproterozoic time just before the Cambrian biologic explosion.

Debate flourishes, however, on whether some climatic events in the Neoproterozoic might have had causes quite different from those in the Phanerozoic. The glacigene–cap carbonate combination, if confirmed and interpreted correctly, may apply to nearly everywhere upon the Earth at the same time (Hoffman et al. 1998a,b; Kennedy et al., 1998). According to this concept, which has become an attractive hypothesis in recent years, as glaciation waxed, worldwide sea ice ensued and quickly modified biogeochemical exchanges between the atmosphere and oceans. Global albedo rapidly increased and the climate was drastically cooled. Sea ice covered global seas and was associated with the waxing of very thin glaciers upon land, glaciers that did not tie up much water to lower sea level. Photosynthetic processes in the upper levels of the sea nearly ceased because of oceanwide sea-ice freezing. For a lapse of ca. 10 m.y. the Earth took on a "snowball" aspect. According to this hypothesis the hydrological cycle was markedly affected. Evaporation sources to feed snow and ice on land were diminished. Attendant coldness reduced river runoff to the sea, and modified the chemistry of weathered and eroded products. The albedo climbed high and the climate cooled drastically. With moisture interchange between the seas and the air nearly shut down, carbon dioxide then built up rapidly in the atmosphere from volcanic sources. Very quickly carbon dioxide and other gases from volcanoes brought about warming, and the worldwide canopy of sea ice and ice on land melted away. Exchanges between the atmosphere and ocean were restored. Circulation of the combined air-ocean system again played its role in heat exchange as a fundamental cause of climate zoning. Questions remain concerning how many times this sequence of events may have happened.

This plausible sequence depends, however, on events to get the glaciation started. Hoffman et al. (1998a,b) consider it likely that the fragmentation of the Rodinia supercontinent created many new continental margins. Moreover, their plate reconstructions place nearly all continents in the Southern Hemisphere so that sea ice probably grew rapidly in a very cold Northern Hemisphere. Perhaps as well, solar luminosity and heat arriving at the Earth were 6–7 % lower than at present (Kasting, 1987, 1992b). Hoffman and colleagues (1998a,b) discount the likelihood that other proposals, such as tipped Earth axis, bolide impacts, and drastic overturn of deep-ocean waters, are satisfactory. In short, geotectonic processes and arrangements and elevations of continents on a mobile crust upon an Earth with a Neoproterozoic biosphere largely dominated by prokaryotic organisms were largely responsible.

On the other hand, Jenkins and Scotese (1998) relate glaciation to collisional events during the amalgamation of Gondwana when there probably were high mountain ranges capped by glaciers with tongues reaching downward to the sea. Hoffman and colleagues consider the northwestern Namibian glacigene strata as laid down on a broad shelf and find no stratal evidence of mountains nearby at that time. Mountains formed ca. 200 m.y. later when Gondwana assembled (Hoffman et al., 1998b). If the region were indeed tropical, weathering may have led to drawdown of carbon dioxide and consequent enhancement of glaciation.

Isotopic analyses from stratal sections here and there over the Earth show sharp δ^{13}C and ^{87}Sr/^{86}Sr excursions (e.g., Kaufman and Knoll, 1995; Kennedy, 1996; Kaufman et al., 1997; Hoffman et al., 1998a,b; Kennedy et al., 1998). These excursions not only are advanced as indicating times of a snowball Earth, but if so, they are viewed as worldwide and synchronous and can be used for interregional correlations. This view is attractive but as yet independent dating of far-apart stratal sections exhibiting the isotopic excursions is lacking. In my view our understanding does not yet allow us to discard all other explanations for the Neoproterozoic icy intervals, and I prefer explanations calling upon a mix of processes that includes those known to have operated in Phanerozoic time.

Nonetheless, if the excursions are shown to be short lived, they provide a most attractive means of worldwide correlation. Kennedy and colleagues (1998) employing cladistic analysis, find that there are two clusters of glaciations, one in late Neoproterozoic time (termed Marinoan) and the other (designated Sturtian). They group the Chuos (Namibia), Rapitan (northwest Canada), and Sturt (Australia = Sturtian) together and the Ghaub (Namibia), Elatina (Australia = Marinoan), and Ice Brook (northwest Canada) together. In contrast, Hoffman et al. (1998a) include the Ghaub with the Sturtian. This disagreement points up the need not

only for further chemostratigraphic analysis but also for other types of investigations. It is also inappropriate to discard from consideration explanations involving a tipped rotational axis or true polar wander as influential, as described above. Incomplete data and interpretations in hand suggest to me that there were several discrete glacial episodes in late Neoproterozoic time, some of which may not be related to the snowball Earth explanation. For example, the arrangement of continental blocks and their influence on air and ocean currents, and the flow of bottom oceanic waters, is as yet unreconstructed satisfactorily for limited time slices. Events and processes that may have contributed to Neoproterozoic glaciation and so far have not been documented include bolide impacts or other extraterrestrial happenings. The challenge remains for geoscientists to find out just how differently the climate system operated in the Neoproterozoic compared with later times although the products of these geological processes seem to be about the same as those operating today. The geological processes, however, were not so different that they prevented the slow evolution of life, although an explosion of biological activity followed beginning in late Neoproterozoic time and accelerating in the Cambrian.

Mid-Proterozoic nonglacial record

Going back in time before the Neoproterozoic, there is no certain record of glaciation on Earth between ca. 950 Ma and 2250 Ma, although glacigene beds were possibly laid down during this interval in Siberia and Scotland (Young, 1991; Eyles, 1993; Eyles and Young, 1994; Davison and Hambrey, 1996, 1997; Stewart, 1997). As yet, however, it is not known whether this gap of ca. 1300 m.y. is real and surface conditions on Earth were significantly warmer *or* more equitable. Did no glaciers thrive and reach down to the sea, or have the vicissitudes of preservation left us with little icy documentation? Mid-Proterozoic glacial deposits may still be discovered and dated satisfactorily, or they may have been eroded away. Enough is now known about crustal rocks of this age, however, to suggest that the nonglacial gap is an interval in Earth history when glaciation was reduced. Perhaps lithospheric tectonics and magmatic processes, and the effluence of greenhouse gases, were somewhat different.

Strata of glacigene aspect have been reported that may have been deposited during this 1300 m.y. mid-Proterozoic interval, but their dating is unsatisfactory. In north-central Siberia, in the Jena River region, Riphean proximal and distal glacial-marine beds grade into terrestrial facies. These are reported from strata assigned to the Nichatka, Bol'shoy Patom, Dzhemkukan, and Choty Formations (Chumakov, 1991; Chumakov and Krasil'nikov, 1992). The glacial units now border the Aldon Shield, Berezovskay Depression, and Chuya Uplift. Diamictite layers are interbedded with stromatolites and microfossils regarded as typically of Middle Riphean age (between 1000 Ma and 1350 Ma). Some investigators, however, place their age within the Late Riphean at ca. 800 Ma so they may be manifestations of evolving Late Proterozoic icy times (Chumakov, 1991). Inasmuch as it is often assumed that glacial episodes were synchronous, glacial horizons may have been placed in the geologic column when glaciation is known elsewhere. Independent dating is required.

In northwestern Scotland glacial paleotopography including roche moutonnée-like forms have recently been recognized at the base of Torridonian strata that are placed insecurely at about 1100 Ma (Davison and Hambrey, 1996, 1997; Stewart, 1997). Overlying diamictites of glacial aspect, rapid lateral facies changes, and especially the occurrence of outsized clasts as dropstones in laminated sandstone facies suggest glaciation; but the age, latitude, and regional setting and extent are largely unknown. Williams and Schmidt (1997b) provide evidence from paleomagnetic studies that paleosols immediately beneath the inferred glacigenic beds developed near 38° lat at ca. 980 Ma. If confirmed this suggested glacial interval would have occurred very early in Neoproterozoic time. The controversial and questionable icy episode is not known to be widespread.

Was the mid-Proterozoic a time of very different tectonic evolution of the crust, different from that prevailing both before and after? Were lithospheric processes different enough to bring on a long interval of warmth so that glaciers nearly disappeared from Earth? The history of crustal evolution during the mid-Proterozoic includes several events and occurrences of unusual rocks that may help our understanding of climates (Cloud, 1976, 1988; Windley, 1984; Hoffman, 1989b; Goodwin, 1991; Schopf, 1992; Lowe et al., 1992; Reed et al., 1993; Trompette, 1994; Canfield and Teske, 1996; Canfield, 1998; Frank and Lyons, 1998; Collerson and Kamber, 1999). The mid-Proterozoic appears to be a time when free oxygen was slowly building up in the system. Knowledge of mid-Proterozoic geologic events and tectonic arrangements remains sketchy, and what is now known shows complexity and repeated overprintings. At times cratons harbored huge intraplate basins, as, for example, the Belt basin in northwestern United States (Link et al., 1993a). Weathering and erosional processes were no doubt different and the diminutive land plants probably consisted mainly of microbial communities (Horodyski and Knauth, 1994).

It seems likely that since some time in the Late Archean, lithospheric history has been characterized by the opening and closing of oceans according to the Wilson Cycle. Cratons have fragmented, and the pieces have drifted apart and then come together again. Such a history, however, is not yet well documented for mid-Proterozoic time over Earth as a whole. Rogers (1996), in looking at what is now known concerning the history of continental accretion and stabilization through time, places the formation of a supercontinent largely between 1600 Ma and 1300 Ma (Gower et al., 1990). The core of this mid-Proterozoic supercontinent may have predominated during the no-glacial-record time gap and included parts of cratonic Siberia, North America, Greenland, and then adjacent East Antarctica. It is deemed to have remained coherent until ca. 1000 Ma when it was incorporated into the supercontinent of Rodinia. In North America most of its central craton was assembled between 1980 Ma and 1650 Ma (Hoffman, 1988, 1989b); western Gondwanan cratons

united at ca. 1250 Ma (Trompette, 1994). Such mobility implies that crustal units moved in and out of climate belts under the premise that the air-ocean system circulated in a way somewhat like that in later time. Until acceptable mid-Proterozoic reconstructions are in hand, it may be that cratonic blocks moved into warm climate belts where there were no widespread glaciers.

On the other hand, lithospheric magmatism and crustal thicknesses may have been different during the mid-Proterozoic gap in the glaciation record. Hoffman (1989b) has proposed that anorogenic magmatism developed beneath a huge supercontinent (probably high-standing), one that aggregated at the end of the Early Proterozoic and just before the putative nonglacial gap. Ponding of mantle melts in the crust is revealed today at places by vast volumes of rhyolite, syenogranite, anorthosite, gabbro, and perhaps other unusual petrologic types such as rapakivi granites (Goodwin, 1991; Mezger et al., 1991; Van Schmus and Bickford, 1993; Aleinikoff et al., 1996; Scoates et al., 1996; Corrigan and Hanmer, 1997). Voluminous outgassings of CO_2 associated with these events may have affected climate, forcing it toward a warmer part of its normal fluctuations (Young, 1991). Or, in contrast, perhaps a high-standing supercontinent lowered sea level, a situation propitious for high ground to harbor glaciers. Moores (1993, 1994; see counter comment by Wise, 1994) noted that sparse data on the preserved thicknesses of ophiolite sequences suggest that oceanic crust manufactured during the time gap is from two to three times thicker previous to 1000 Ma than it is later. He argues that this situation brought about very high sea levels that flooded continents so that they were from 90% to 95% inundated. Other unusual products are reported during the interval, such as "molar-tooth" structures implying different geochemical processes. A worldwide uniform temperature is viewed as the consequence.

The effect of bolide impacts and extraterrestrial dust upon climate during the mid-Proterozoic gap also needs appraisal, but there is no clear evidence that it was a time of increased impacting (Shoemaker and Wolfe, 1986; Shoemaker et al., 1990). A tabulation by Grieve et al. (1995) lists several bolides with craters that may have impacted during the interval although the dating on several is quite approximate: Iso-Naakkima (Finland, >1000 Ma), Kelly West (Northern Territory, Australia, >550 Ma), Lawn Hill (Queensland, Australia, >515 Ma), Lockne (Sweden, >455 Ma), Lumparn (Finland, ca. 1000 Ma), Spider (Western Australia, >570 Ma), Sudbury (Ontario, Canada, 1850 ± 3 Ma), Teague (Western Australia, 1630 ± 5 Ma), Tvären (Sweden, >455 Ma), and Vredefort (South Africa, 2006 ± 9 Ma).

Future study is required before mid-Proterozoic tectonic and magmatic and bolide events can be related to climate change. Paleomagnetic apparent polar wander paths for North America show large displacements suggesting that craton rearrangements took place that are bound to affect climate (Elston et al., 1993). During the interval environmental stability is inferred from the record in Australia (Brasier and Lindsay, 1998) whereas supercontinental fragmentation is suggested in Zimbabwe (Oliver et al., 1998). Satisfactory geotectonic explanations for the gap remain elusive; further discussion is beyond the scope of this review of the glacial record. Only speculations are now at hand on whether events enhanced or reduced carbon dioxide fluxes into the climate system; whether sea level rose or fell; what was the effect of evolving life; and whether sea and land arrangements affected water vapor distribution, albedo, and pathways of evaporation and precipitation along circulation belts. Life continued to evolve through the interval (Schopf and Klein, 1992; Horodyski and Knauth, 1994), however; previously, there was certain glaciation, so it is doubtful that climate departed very far from that prevailing both before and after.

EARLY PROTEROZOIC ICE AGES

Between ca. 2200 and 2400 Ma continental glaciation is recorded within three far separate regions: in Canada and adjacent United States, in South Africa, and near the Finland-Russia border (Young, 1968, 1970, 1983, 1984, 1988, 1991, 1992a; Young and Nesbitt, 1985, 1999; Mustard and Donaldson, 1987; Visser, 1971, 1981; Marmo and Ojakangas, 1984). The sparse and piecemeal glacial record is convincing but the dating is insecure so that it is not known whether these now far-separated glaciations were synchronous, whether they were separate episodes, or whether the sites were originally quite close together. In both Canada and South Africa, there is strong evidence of low-latitude deposition of glacigene beds from paleomagnetic studies (Evans et al., 1997; Williams and Schmidt, 1997a). In addition, in Western Australia, mountain glaciation is shown by glacially faceted and striated stones in diamictites and associated strata with an age now known only as lying somewhere between 2000 Ma and 2500 Ma.

North America: Huronian glaciations

Lower Proterozoic strata in southern Ontario, Canada, and especially along the north shore of Lake Huron, include several formations with a glacial imprint. The Gowganda Formation, for example, has long been recognized as of glacial origin (Coleman, 1907, 1926), and modern studies have now elucidated the sedimentary facies and depositional environment of this and other formations (Young, 1981a, 1983, 1988; Miall, 1983, 1985; Mustard and Donaldson, 1987; Ojakangas, 1985; 1988; Young and Nesbitt, 1999). Here the Paleoproterozoic sequence (Huronian Supergroup) lies unconformably upon Archean basement and has been subdivided into several groups and formations. The age of glaciations is bracketed between 2670 Ma, the U-Pb age of felsic volcanic rock at the base of the succession (Krogh et al., 1984), and 2220 Ma, the age of the crosscutting Nipissing diabase (Van Schmus, 1965, 1980; Van Schmus and Hinze, 1985; Corfu and Andrews, 1986; Van Schmus et al., 1993). The timing of the glaciations within this 450 m.y. duration, however—nearly as long as the total span of Phanerozoic time—is uncertain. A role for glaciation has also been advocated for formations within the Huronian below the glacial Gowganda, including the Ramsay

Lake and Bruce Formations (Young, 1981b, 1988). Iciness is also recorded in several other places in North America at distance from the Gowganda beds.

The evidence for glaciation in the Gowganda consists of coarse and sparse-stoned diamictites lying directly on glacially striated pavements (Schenk, 1965; Young, 1981a), faceted and striated megaclasts, thin-bedded sequences with undoubted dropstones and other sedimentary structures that are characteristic but not necessarily diagnostic of glacial and periglacial environments (Fig. 65). Clasts within the formation come from many sources, some of which are viewed as extrabasinal in provenance. Some of the diamictites are probably lodgment and ablation tills from continental ice sheets (Harker and Giegengack, 1989). Many deposits resulted from debris flows that carried tillitic and other material down slopes into deeper water (Lindsey, 1971; Mustard and Donaldson, 1987; Miall, 1983, 1985). The Gowganda is widespread over a region at least 400 km wide in an east-west direction, and 200 km in a north-south direction.

Three upward-fining sequences are recognized that begin with diamictite at the base: Ramsay Lake, Bruce, and Gowganda Formations. The older diamictites and related rocks, assigned to the Ramsay Lake and Bruce Formations of the Lower Huronian, were laid down in a restricted locale. Striated and faceted stones from within these two formations have not been recognized. The similarity of these formations with the overlying Gowganda, however, lends credence to a glacial origin for these deposits as well.

Reconstructions of the paleogeography and tectonic setting of the Gowganda and related strata by Young (1988) suggest a marine environment at a foundering continental margin. The Lower Huronian deposits, less widely distributed than those above, are viewed as laid down in a rift associated with extensional faulting that evolved into a continental margin that in turn developed into a passive margin. These lower beds contain some terrestrial deposits, including fluvial sandstones (Long, 1973,

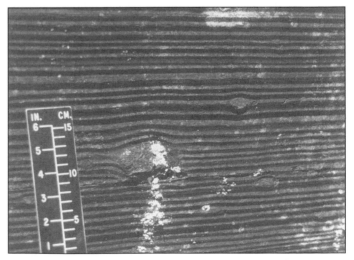

Figure 65. Dropstone pebble in laminated graywacke-argillite sequence. Paleoproterozoic Gowganda Formation, north shore of Lake Huron, near intersection of Highways 129 and 554, Canada. October 1985.

1981). Following the rifting, Gowganda glaciers spread widely across the continent itself with lobes extending seaward. Tillitic debris moved downslope into deeper water. After the inferred continental breakup with associated foundering, the continental margin subsided, glaciation ended, and seas encroached upon the continent. Interpretations of lutite geochemistry within the sequence, especially from the Serpent Formation immediately underlying the Gowganda Formation, suggest continental weathering processes compatible with rapid cratonic drift from tropical into cold regions (Nesbitt and Young, 1982; Fedo et al., 1995, 1997; Young and Nesbitt, 1999).

Glacigene strata in part correlative with the Huronian strata are recognized elsewhere in central and northern North America. Along the margin of the Superior craton northeastward from Lake Huron across Quebec as least as far as the Chibougamau area, Huronian rocks occur (Lindsey, 1969; Long, 1981; Young, 1988). To the west of the Lake Superior region, the base of the Marquette Range Supergroup lies unconformably upon basement and probably correlates with the Gowganda (Gair, 1981). In the Canadian Northwest Territories, basal beds of the Hurwitz Group (Padlei Formation) may also represent deposition during waning glaciation at about the same time. Here ice rafting is indicated by dropstones in mudstone layers associated with diamictites (Young, 1970, 1973; Young and McLennan, 1981). A gabbro sill within the Hurwitz Group ~1,200 m stratigraphically above the inferred glacigene beds of the Padlei Formation gives a U-Pb zircon-baddeleyite age (2094^{+26}_{-17} Ma) that establishes a minimum age of ca. 2100 Ma (Patterson and Heaman, 1991). These Hurwitz beds lie more than 1,500 km north of Huronian strata in southern Ontario, and therefore may not be part of the same tectonic province.

South of Lake Superior in northern Michigan ~200 km west of the type Huronian sequences, two formations contain a glacial imprint: the Fern Creek and Reany Creek Formations (Pettijohn, 1943; Puffett, 1969; Young, 1970, 1981b, 1988; Gair, 1981). They were probably laid down in a now-western part of the same basin as the Gowganda and other units of the Huronian Supergroup. The Fern Creek beds were in part deposited unconformably in hollows and swales upon the basement surface and include dropstones in laminated argillites associated with polymictic metaconglomerates and breccias containing boulders and blocks up to 1.5 m in length. The Fern Creek is correlated with the Reany Creek metaconglomerates, metagraywackes, and slates lying ~90 km to the north, and both of these probable glacigene units in turn with the Gowganda ~250 km to the east (Ojakangas, 1988; Young, 1988). Although this correlation is permitted by the sparse radiometric dating, complex tectonics, overprinting, and metamorphism preclude assured correlation.

In southern Wyoming (Medicine Bow Mountains) there are probable glacigene deposits, assigned to the Snowy Pass Supergroup (Headquarters Formation) and are now deformed and metamorphosed (Houston et al., 1981, 1992; Houston, 1993; Houston and Karlstrom, 1993). Diamictites within the succession contain granitic and metamorphic boulders, rounded to subangular, and

up to 1.5 m in size. The megaclasts make up as much as 20% of the diamictites embedded in a sandy matrix, now metamorphosed. Some faceted stones and some with striations of possible glacial origin are reported, and dropstones within laminated phyllites are associated with the metadiamictites. Limited paleocurrent studies suggest a northeastern source that provided debris to both a nonmarine and marine environment. Regional relations place the age between 2700 Ma (the age of the Archean basement) and 1700 Ma (the age assigned to overprinting metamorphism). Within this 1000 m.y. interval, Houston et al. (1981; Houston and Karlstrom, 1993) tentatively suggest an age of between 2700 Ma and 2100 Ma because of the lithological resemblance to the Huronian strata of the Lake Huron area. Although this correlation is attractive, it contains an element of "circularity" in reasoning. In addition to questions concerning timing, the two regions are now nearly 2,000 km apart and their tectonic settings originally may have been either closer or farther apart. The Wyoming strata lie above the Wyoming Archean Province, whereas those of the Huronian lie above the Superior Archean Province (Hoffman, 1989b). These ancient cratons are separated by the Trans-Hudson orogenic belt, which originated with rifting at ca. 2100 Ma and culminated with left-oblique collision at ca. 1800 Ma. The broad tectonic zone between the Huronian and Snowy Pass glacigene strata, including younger beds, may have opened and then closed, but they may have been adjacent before the opening of this presumed ancient ocean. Roscoe and Card (1993) suggest that both the Canadian Huronian and Wyoming Snowy Peak strata were deposited in a single epicratonic basin that was separated following upon widespread gabbro intrusions at ca. 2200 Ma and then reassembled at ca. 1850 Ma. Rift shoulders during some of these events may have provided attractive sites for the initial accumulation of ice.

Finland-Russia border region: Karelian glaciations

Glacigenic strata are preserved in the Karelian region of eastern Finland and adjoining Russia with an age between 2450 Ma and 2180 Ma—a 270 m.y. interval (Negrutsa and Negrutsa, 1981; Marmo and Ojakangas, 1984; Ojakangas, 1988). In Finland the glacigene strata constitute the Urkkavaara Formation, and in adjacent Russia several glacigene units are placed within the Sariolian Group.

The metasedimentary Urkkavaara Formation in Finland consists of four informal members: a lower metasiltstone-argillite (more than 15 m thick), overlain by a graded metasandstone member (10–15 m thick), in turn overlain by the upper metasiltstone-argillite member (2–40 m thick), and topped by a metadiamictite unit (1–10 m thick) (Marmo and Ojakangas, 1984). Dropstones are replete in the metasiltstone-argillite units, now schists, that consist of laminated metasiltstone embedded in argillite. Oversized dropstones as large as 20 cm are composed of felsic plutonic rocks and elongated phyllitic clasts up to 10 cm. Many of the dropstones have either pierced or bent laminations during deposition or have bent them during compaction, deformation, and metamorphism. No faceted or striated stones were found despite careful search, but in such deformed and metamorphosed strata their preservation is unlikely. Aggregates of sand, silt, and clay (now sand-silt-mica clusters) are interpreted as metamorphosed till pellets dropped as icebergs or ice floes melted or overturned. The massive sparse-stoned metadiamictites are interpreted as formed by "rainout" of clay and silt (rock flour?) into a quiet environment lacking lateral currents for sorting. The gradational boundary at the base of the metadiamictite unit argues against downslope sliding, which would normally result in a scoured or erosional lower contact. Rare laminated metasiltstone-argillite slabs with chaotic internal structure within metadiamictite are viewed as the result of local slumping in areas of rapid sedimentation. Graded beds imply deposition by turbidity currents in a glacial-marine or glacial-lacustrine belt offshore from glaciers themselves. Rare lonestones in the turbidites are also interpreted as dropstones by Marmo and Ojakangas (1984) rather than as lagstones left behind as a vigorous turbidity current passed by. Ojakangas (1991) notes that the Finnish glacial sequences are overlain by a metaregolith with a composition suggesting deep weathering in a subtropical to tropical climate. To explain this, he favors tectonic movement of the glaciated continental sites from relatively high latitude to low latitude as suggested by rapid drifts and rotations of Fennoscandian cratonic plates shown by sparse paleomagnetic data (Pesonen et al., 1989). The Urkkavaara Formation has been traced for at least 300 km in Finland.

To the east of Finland, across the border in Russia, similar units within the Sariolian Group are also of probable glacigene origin (Negrutsa and Negrutsa, 1981). Lenticular metadiamictite bodies lie at several stratigraphic levels interbedded with metavolcanic units. These are also associated with laminated strata including banded schists with dropstones (Salop, 1983; Ojakangas, 1988). Stones within the metadiamictites, constituting about 5%, do not exceed 8 cm in size, and no striated stones have been reported. These metadiamictites apparently occur in about the same stratigraphic position as those of the Urkkavaara Formation in Finland.

The Karelian belt of metadiamictites and associated rocks extends over 600 km in separate patches within Russia parallel to the Finnish-Russian border. Near Lake Ladoga on the south, metadiamictites occur in beds of the Ladoga Group (Yanis-Yarvi tilloids) that apparently fit within the upper part of the Karelian sequence (Negrutsa and Negrutsa, 1981). Similar metadiamictites occur northwest of Murmansk (Lammos tilloids), ~500 km to the north of the Urkkavaara localities, but little evidence suggesting a glacial origin has yet come to light for either of these Russian regions (Negrutsa and Negrutsa, 1981). Correlation is uncertain because the Lammos metadiamictites may lie stratigraphically well above the Urkkavaara horizon, and north of a broad tectonic belt (Lapland-Belmorian Zone of younger Proterozoic age) that separates the Karelian Massif on the south from the Kola Massif on the north (Salop, 1983; Goodwin, 1991).

Reconstructions of the paleogeography during the time of the Karelian glaciations, based largely on sparse paleocurrent data in units overlying the Urkkavaara glaciogene beds, suggest current flow in a now-westerly direction from a subdued granitic continent (Marmo and Ojakangas, 1984). The overlying meta-arkoses and metaconglomerates are viewed as deposited as sea level rose after a regional glaciation, which may have been at low latitude according to poor paleomagnetic data (Hambrey and Harland, 1981). The evidence for glaciation upon a low-lying continental shelf adjoining a craton at this time is reasonably convincing for this region but the dating is insecure. Pb-Pb isochron dates give an age for underlying basement at 2455 ± 45 Ma, and 2180 ± 60 Ma for metavolcanic rocks in the upper part of the sequence well above the glacigene rocks (Negrutsa and Negrutsa, 1981). This span of 230 m.y. overlaps that of the Huronian glaciations in North America (2180–2670 Ma), but it is uncertain whether the two now-distant glacial episodes occurred at the same time within this long interval. Data now in hand indicate broad correlation but only within a long time span; ample time remained for diachronous glaciation as cratons carrying glacial imprints moved about.

Southern Africa, Griqualand West and Transvaal regions: Makganyene glaciation

In South Africa and adjoining Botswana, Paleoproterozoic strata with glacial affinities are present in the Griqualand West and Transvaal Supergroups (Visser, 1971, 1981; De Villiers and Visser, 1977; Tankard et al., 1982; Powell and Neumiller, 1986; Evans et al., 1997; Evans, 1998). The glacigene beds in these two stratal sequences extend in a belt southeasterly from the Pretoria region for ~700 km. They occur in two regions now separated by at least 300 km as the result of postdepositional uplift and erosion but are interpreted as laid down within a single elongate basin. In the Griqualand West region on the southwest, the glacial record is discontinuously preserved in the Makganyene Diamictite Formation near the base of the Postmasburg Group. Rhythmically laminated shale contains small dropstones and diamictite layers that consist of about 6% by volume of stones, commonly cobbles and pebbles. These are interbedded with argillaceous sandstone, shale, and conglomerate, and at places overlie banded iron formation and carbonate beds. Du Toit (1954) mentions a boulder of ~40 cm, and Visser (1981) reports faceted and striated stones. Some diamictite layers contain wisps and structures suggesting downslope sliding, whereas other facies imply deposition near a glacial front (Powell and Neumiller, 1986).

In the Transvaal basin ~500 km to the northeast of the Griqualand West region, the glacigene beds are assigned to the Timeball Hill Formation (Pretoria Group) consisting of diamictite, shale, siltstone, and sandstone. Visser (1981) reports an angular megaclast with crescentic fractures and gouges and N. J. Beukes and I collected several glacially faceted and striated pebbles from a roadside quarry near Magaliesburg, ~75 km west of Johannesburg on July 19, 1994. One flat stone had apparently fallen through the water column and penetrated laminations at a high angle. We inferred that ice rafting had taken place in an aqueous environment, probably marine. Structures within the diamictite layers, intercalated with thin-bedded and laminated sequences, display the results of some sliding along a gentle slope. The sequences range in thickness between 80 m and 530 m over a distance of ~500 km.

Convincing low-latitude paleomagnetic measurements of $11 \pm 5°$ have recently been obtained from the Ongeluk Lava exposed in the central part of the Griqualand West region in the vicinity of Postmasburg (Evans et al., 1997). They come from lavas lying conformably above the Makganyene, and give a Pb-Pb age of $2,222 \pm 13$ Ma. Fragments of volcanic rocks of the same type occur within the Makganyene, and conformable contacts between the glacigene formation below and the lavas above imply that there was no long time lapse between the extrusion of the dated lavas and the glaciation. An ion-microprobe U-Pb date of 2552 ± 11 Ma is reported from a tuff in the sequence unconformably below the Makganyene (Barton et al., 1994). Lavas within the Transvaal basin, stratigraphically above the glacigene strata, give an age of 2224 ± 21 Ma using several isotopic methods (Walraven et al., 1990); other results from a recalculated Rb-Sr age give 2177 ± 21 Ma (Steiger and Jäger, 1977). These dated layers lie stratigraphically well above the basal units of the Transvaal Supergroup, from which several dates of ca. 2350 Ma have been obtained. Under the premise that these correlations across more than 500 km are correct, and that glaciation was roughly synchronous in this region, an ice age is here placed at ca. 2250 Ma. Glaciers probably lay upon a stable platform at the margins of the West Griqualand–Transvaal basins on the now-southeast (Visser, 1981), but regional paleogeographic reconstructions remain incomplete.

Western Australia: Hamersley glaciers

Diamictites containing glacially striated stones occur along the margin of the Hamersley basin, Western Australia, assigned to the Meteor Bore Formation (Turee Creek Group) and document glaciers somewhere upslope not far away (Trendall, 1976, 1981). The metamorphosed and cleaved strata are interpreted as laid down in a fluvially dominated environment, probably in a foreland basin near a convergent continental margin, and reach a thickness of 300 m (Goddard, 1992). Bouldery metadiamictite is interbedded with fine-grained metagraywacke and metasandstone. Faceted and striated stones, well illustrated in Trendall (1976), are convincingly of glacial origin. Stones within the diamictites are matrix supported, but rip-up and flow structures have not been reported to indicate definitely a debris-flow origin. Bedding is obscure except where there are rare intercalations of dolomite up to 1 m thick. No dropstones have been reported. It seems likely that the glaciers were sited within mountains bordering the basin and not from a widespread ice cap with ice tongues reaching the sea, but this interpretation is uncertain. As yet, no useful paleomagnetic data to indicate a latitude of deposition for the deposits have been reported.

The age of the Meteorite Bore unit is poorly constrained

between ca. 2000 Ma and 2500 Ma (Trendall, 1981). Deposition within the Hamersley basin may have begun as early as 2700 Ma and lasted until 1850 Ma based on sparse isotopic data and stratigraphic interpretations. The glaciation at the basin margin may have occurred at roughly the same time as other Paleoproterozoic glaciations over the world but this is by no means assured. A counterargument comes from the observation that some of the volcanic stones within the unit resemble those of the Woongarra Volcanics, the only known local source for such rocks, and these are given an age of ca. 2000 Ma. If this correlation is confirmed, the Meteorite Bore Member and the glaciation are younger than 2000 Ma and at least 300 Ma younger than the current best estimates of the ages of the other Paleoproterozoic glacigene sequences.

Discussion

Although the stratigraphic record confirms that there were strong glaciations in three or four now far-separated parts of the globe in Paleoproterozoic time, several questions arise: Were these glacial episodes truly synchronous, or were they diachronous within a long time span when the climate was cooler than average? The dating for *each* glaciated region cannot be narrowed down in time to less than ca. 300 m.y. (Fig. 66). If, however, time spans are narrowed by estimates of where within a stratigraphic pile the glacial record occurs, and if it is assumed that approximate synchroneity is warranted, there is a common interval between 2325 m.y. and 2245 m.y. a span of ca. 80 m.y., widened here to 100 m.y. (2350–2250 m.y.) to emphasize the uncertainties. Far better dating is needed before this speculation is acceptable. There may or may not be a time-restricted "megaevent" that can be employed for correlation (Ojakangas, 1988). In addition, could the three main areas have been grouped together ca. 2000 m.y. ago and, since then, drifted apart?

Mobility of cratonic blocks according to the Wilson cycle with the fragmentation of cratons, the opening and closing of scaways, and the rising of collisional orogenic belts is an acceptable scenario for Paleoproterozoic time (e.g., Hoffman, 1988, 1989a,b, 1991; Roscoe and Card, 1993; Rogers, 1996; R. G. Park, 1997). Such a view is supported by tectonic reconstructions, interpretations of stratigraphic sequences (e.g., Young, 1995a,b), and chemical weathering studies (Holland and Beukes, 1990; Young and Nesbitt, 1999). Reliable paleomagnetic data on the latitude when glacigene Paleoproterozoic beds were deposited are lacking except for the Makganyene deposits (Evans et al., 1997). Here low latitudes of $11 \pm 5°$ at ca. 2350 Ma are deemed acceptable. High latitudes along with rapid movement of cratons from high to low latitudes for the Karelian glacial record in Russia are suggested by sparse paleomagnetic data and an immediately overlying megaregolith implying subtropical or tropical weathering (Hambrey and Harland, 1981; Negrutsa and Negrutsa, 1981; Pesonen et al., 1989). Latitudes between 38° and 66° are suggested for the Huronian strata in southern and eastern Ontario (Young, 1981a; Long, 1981).

During Paleoproterozoic time oxygen levels were increasing

and carbon dioxide levels decreasing as shown by paleosol, isotopic, and other studies (e.g., Walker, 1977; Holland et al., 1986; Burdette et al., 1990; Schopf and Klein, 1992; Kasting, 1993; Karhu and Holland, 1996; Ohmoto, 1996; Gutzmer and Beukes, 1998). It was a time when biogeochemical processes increased their influence on the climate system. With cratonic mobility and presumed reduction in carbon dioxide, sites for glaciation probably moved in and out of appropriate latitudinal positions.

ARCHEAN GLACIATION

South Africa

The earliest glaciation so far recognized on Earth is recorded within strata of the Witwatersrand and Pongola basins of South Africa (Fig. 67) (Du Toit, 1954; Wiebols, 1955; Haughton, 1963, 1969; Harland, 1981; Tankard et al., 1982; Crowell, 1983b; Von Brunn and Gold, 1993; Gold and Von Veh, 1995; Young et al., 1998). Modern zircon dating indicates that the glaciation

occurred between 2990 Ma and 2914 Ma (Robb et al., 1991; de Wit et al., 1992) in early Late Archean (Ar_3) time (Palmer, 1983; Harland et al., 1990). Archean time is conventionally agreed upon as ending at 2500 Ma (Plumb, 1991), and the latest of three subdivisions (Ar_1, Ar_2, and Ar_3) is shown in charts as beginning at 3000 Ma, only a bit older than the glacigene strata discussed here. The evidence for glaciation consists of glacially faceted and striated stones occurring within outcropping diamictite layers, upon the interpretation of depositional sites of diamictites and associated beds, and upon interpretations of the origin of underground metadiamictites exposed in mine workings.

In Witwatersrand strata of the Klerksdorp area of the West Rand region, 170 km southwest of Johannesburg, several layers of diamictite crop out. They occur in the Government Subgroup, (West Rand Group), ~2,800 m above its basal contact with the Dominion Group (Tankard et al., 1982). Clasts of quartz, quartzite, and chert within the deeply weathered diamictite exceed 40 cm in diameter, and some of those recovered more than 20 cm in diameter are glacially faceted and striated. The stones

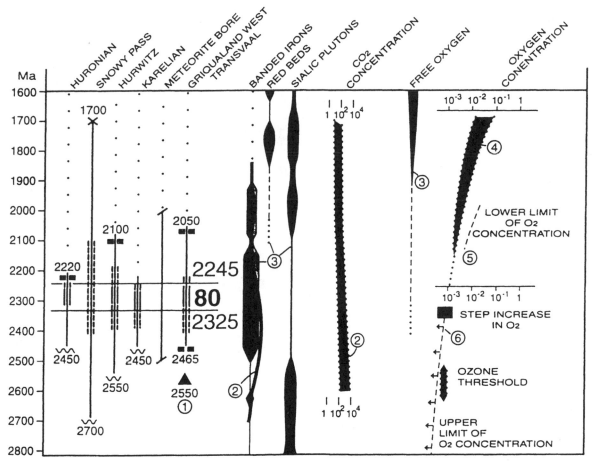

Figure 66. Paleoproterozoic time chart of glacial and other events, by regions discussed in text. Ranges of ages depicted by single bar, but likely ages by triple bars. Under the unsupported premise that the far-flung evidences of iciness were approximately synchronous, there may be an 80 m.y. interval between 2245 Ma and 2325 Ma when glaciation may have occurred. Data from (1) Barton et al. (1994); (2) Klein and Beukes (1992); (3) Cloud (1988); (4) Schopf (1993); (5) Runnegar (1991); Schopf (1992, fig. 13.1); (6) Walker et al. (1983).

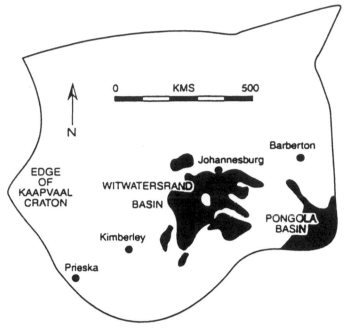

Figure 67. Location of Witwatersrand and Pongola basins, South Africa, where record of nearby Late Archean glaciation is recorded. After De Wit et al. (1992, fig. 1b).

above those of the Black Bar. Most of the pebbles consist of white quartz along with chert, quartzite, indurated shale, and some dark fine-grained igneous rocks. Inasmuch as the metadiamictite is indurated, stones cannot be removed easily for examination so no faceted or striated stones were recovered and examined. Even if facets and striations were originally present, however, they probably have been destroyed by metamorphism.

I deem it likely that some of the material within the Witwatersrand indurated diamictites observed below ground, as well as those that crop out, were in part derived from glacial debris in the source area. The occasional occurrence of glacially polished, faceted, and striated clasts, such as those found in weathered outcrops where the stones come out whole so that their surfaces can be examined, suggests local glaciation somewhere at basin margins contributing to the debris that then moved downslope in channels.

are embedded in a matrix of poorly sorted sandy siltstone with massive structure (Du Toit, 1954; Wiebols, 1955). North of Klerksdorp within the "Lower tillite," with other geologists I examined several glacially faceted and striated cobbles (August 20, 1982). One in particular, 5 × 10 cm across, displays convincing and photographable glacial striations on a flat facet (Fig. 68) (Crowell, 1983b). Similar clasts have been recovered previously by several geologists.

Approximately 25 km east of Johannesburg, near Boksburg, Martin et al. (1989) interpret diamictites exposed in mine workings, assigned to the Black Bar, Main Conglomerate Formation, as deposited by mudflows derived from a tectonically active region at the basin margin. The subsurface diamictites lie nearly 2,000 m stratigraphically above those of the Government Subgroup mentioned above although there is much lateral thickness change across the region so that thickness estimates are quite approximate. These investigators find no need to call upon glaciation for the Black Bar debris before it moved downslope through channels into the basin.

In underground workings ~25 km farther southeast from Boksburg (and 200 km east of the Klerksdorp locality), I have briefly examined metadiamictite, aptly termed "snowstorm conglomerate" (also "puddingstone," "tilloid," and "tillite") occurring in broad channels at least 100 m wide and 4 m deep that are cut into bedded and cross-bedded quartzite (August 19, 1982) (Number 3 Shaft, Marievale Consolidated Mines, Ltd., East Rand). These metadiamictite channels lie within the Kimberley Reef (reef = conglomerate) ~3,000 m stratigraphically above the diamictites assigned to the Government Subgroup and ~1,000 m

Figure 68. Late Archean glacially faceted and striated stone from the "Lower tillite," Government Subgroup, Witwatersrand Supergroup. Between ca. 2914 Ma and 2990 Ma in age. From Buffelsdoorn Farm, ~20 km north of Klerksdorp, South Africa. Coin is ~1.5 cm in diameter. August 1982.

Published studies of the tectonic and sedimentation history of the Witwatersrand basin tentatively indicate that the part of the section including the diamictites considered here was deposited during a tectonic transition between an earlier time of regional extension, 2900–3100 Ma, and regional shortening, 2700–2900 Ma (Martin et al., 1989; Myers et al., 1990; Stanistreet and McCarthy, 1991; Robb et al., 1991; de Wit et al., 1992). Reconstructions suggest a passive margin on the present north during the extensional stage, with open ocean on the south. The transition stage was followed by a foreland basin stage associated with regional compression which is why I suggest that local glaciers may have occurred around basin margins. During this tectonic transition, the basin shrank and synsedimentary block faulting and associated folding began. Stanistreet and McCarthy (1991), in interpreting the complex structure and sedimentary record, suggest that the basin was "squeezed out" to the southeast. Source areas for the sediments flowing into the basin lay around its deforming margins, and it seems likely to me that glaciers contributed debris but it is not yet known whether these hypothesized glaciers were alpine, piedmont, or ice caps. Weathered and eroded debris from the source areas is viewed as having been washed across an alluvial or outwash apron and into the Witwatersrand basin at distance. The transported debris occurs over a region that is now more than 200 km in breadth and contained within a basin that is inferred to have been ~800 km wide originally (de Wit et al., 1992).

Strata within the Pongola sequence to the east of the Witwatersrand localities contain a record of a glacial contribution from sources an unknown distance from the site (Von Brunn and Gold, 1993; Gold and Von Veh, 1995; Young et al., 1998). The section laid down in the Pongola-Mozaan basin is especially well exposed near the Natal-Swaziland border downhill from the Klipwal Gold Mine. Here I briefly examined diamictites and interbedded strata along the Mageza River, a tributary of the Pongola River, guided by Victor von Brunn (July 23, 1994). Convincing glacially faceted and striated cobbles and boulders are supported in massive sandy and muddy diamictites. A glacial contribution from basin margins seems required. Lithostratigraphic and sedimentological studies and isotopic dating show that the section satisfactorily correlates with similar units in the Witwatersrand basin (Beukes and Cairncross, 1991). Single crystal U-Pb methods, supplemented by other techniques, place the age older than 2870 Ma and younger than 2940 Ma. Deposition is viewed as having taken place on a shallow marine shelf associated with a fluvial braided plain under the influence of sea-level changes.

Are there other Archean glacial records?

A clear record of glaciation as ancient as the early Late Archean deposits in South Africa has not been documented elsewhere upon Earth. Mid-Archean metadiamictites in Montana occur stratigraphically below the Stillwater Complex, a complex consisting of peridotite, norite, gabbro, anorthosite, and related rocks (Page, 1977, 1981; Page and Koski, 1973).

Deposition of this sequence took place sometime between 3140 Ma and 2750 Ma, a 390 m.y. interval (Nunes and Tilton, 1971). Dispersed blocks of diverse shapes up to several meters across are embedded in a dark fine-grained matrix in beds as much as 60 m thick that are traceable for over 10 km. Despite complex structure and hornfels metamorphism, relict sedimentary structures, such as wavy cross-bedding, pinch-and-swell, and soft-sediment deformation structures, occur in the metasedimentary sequence (Page and Koski, 1973; N. Christie-Blick, personal communication, 1980). Unequivocal dropstones have not been reported. Overprinting deformation and metamorphism have obliterated any glacial facets or striations if there ever were any. Downslope sliding in a tectonically active region provides an acceptable explanation but whether glaciation contributed to the debris is unknown. Note, however, that the range in age of the Stillwater strata (3120–2750 Ma) includes the range of the Witwatersrand-Pongola beds (2990–2914 Ma) so the deposits may be roughly contemporaneous.

Archean metadiamictites have also been reported from several stratigraphic levels in Karelia (northwesternmost Russia) and along the eastern margin of the nearby Baltic Shield (Negrutsa and Negrutsa, 1981). Subangular to rounded stones up to 8 cm in size embedded in sandy diamictitic schists consist largely of granite, gneiss, and quartzite, and are deemed to have had local origins. Granitoid rocks cutting the sequence give an age for it older than 2750 ± 59 Ma based on U-Th determinations on zircons, and the basement unconformably beneath the sequence tentatively has a K-Ar age of 3400 Ma. Metamorphosed marly layers, volcanic horizons, and irregularly rhythmic beds are also recognized. The depositional site is inferred to have been near a tectonically active margin of a sea. No evidence for glaciation has been reported.

In summary, it seems likely that glaciers, at least in upland areas near the sea, existed on Earth in Late Archean time. An interpretable record from sedimentary rocks so old, so poorly preserved, and so sparsely distributed is indeed remarkable. I interpret the vestige recovered from Witswatersrand and Pongola strata as therefore of special significance and conclude that at least mountain glaciers existed on Earth 3000 m.y. ago.

Archean-Proterozoic transition

Beginning sometime in the Early Archean, a long period of transition lasting for more than 500 m.y. is visualized when small and discrete continental blocks grew in size as the result of accretion and consolidation of continental material to form protocontinents (Cloud, 1988). These protocontinents or incipient cratons probably moved about as convective processes operated, similar to those sea-floor spreading processes we see today, but perhaps operating more vigorously. Island arcs grew and many were accreted to the enlarging protocontinents. By the mid- or Late Archean and continuing into the Proterozoic, some protocontinents had grown big enough to constitute cratons and influence the style of tectonic evolution. Subduction zones along their margins gave rise to accretionary belts with fore-arc basins, arcs, orogenic belts, and foredeeps. Disorganized tectonic mobility prevailing beneath the early ocean floor was becoming organized. One of the earliest of these evolving cratons was the Kaapvaal craton of southern Africa that began to assemble after ca. 3700 Ma (de Wit et al., 1992) and others were gathered together thereafter (Hofmann, 1987). Granitic intrusions and their volcanic counterparts began to invade and thicken the crust, leading to consolidation of cratons. Epicontinental geographies followed, accompanied by sedimentation into continental basins. True continents were at last assembled upon the Earth and the Wilson cycle began. Perhaps this hypothetical scenario allows for a decrease in the tectonic overturn along early mid-ocean ridges and a corresponding decrease in the effluence of carbon dioxide to the atmosphere and oceans. Perhaps some mountain ranges or rift margins near the sea were high enough to gather snow and ice. Continents were at last at hand by the mid-Archean to provide sites for glaciers, first upon mountain ranges and then to lowlands with tongues reaching to the ancient seas. As the continents enlarged by amalgamation and continued accretion and the Wilson cycle evolved, there came a time when continental fragmentation ensued. Rifts cut across portions of the early continents and their high marginal shoulders provided sites for glaciers. Convergent continental margins are viewed as ringed with ranges, and some of these may have harbored glaciers upon their heights.

Complex interplays affected the oceans and air, but their relative balances are not yet elucidated (e.g., Schopf and Klein, 1992). Life existed and evolution was underway. Photosynthesis and biological activity began to contribute significant oxygen to the ocean-atmosphere system. The Wilson cycle was launched and the Earth had entered a long period of tectonic activity with a style somewhat similar to that prevailing today, but viewed as characterized by significant differences in scale and speed and regularity. Unfortunately, the record is too sparse and piecemeal to find documentation for these speculations.

Many authors, beginning with Svante Arrhenius about 1896 and continuing with Chamberlin in 1899 (Raymo, 1991; Fleming, 1992, 1998; Weart, 1997) have pointed out that a decrease in the proportion of greenhouse gases in the atmosphere (water vapor, CO_2, and methane primarily) would lead to cooling. The role of changes in water vapor content in the atmosphere at these times is quite speculative because reconstruction of the positions of open seas and continents with respect to climate belts is unknown. And so are the effects of changes in albedo. Early in its history the Earth's atmosphere was probably rich in CO_2—probably much richer than now—but along with the evolution of life, the composition of the ocean and atmosphere changed (Cloud, 1968, 1988; Walker et al., 1983; Schopf, 1982; Schopf and Klein, 1992; Kasting, 1993). Several studies—geochemical, biological, and computer modeling—suggest that the partial pressure of O_2 increased beginning at ca. 2300 Ma, primarily due to the rise in photosynthesis (Awramik, 1992). The atmosphere previously was rich enough in CO^2 that the Earth was warm, and widespread

continental glaciation was perhaps precluded except for mountain glaciers nearly 3000 m.y. ago.

Astronomers hold the view that the heat output of the sun has been increasing slowly and steadily so that in Archean times the heat received on Earth was significantly less—the faint young sun hypothesis (Kasting and Ackerman, 1986; Kasting, 1992a,b, 1993; Caldeira and Kasting, 1992). This hypothesis suggests that the climate was cooler in the remote past and that solar insolation has slowly brought about some warming on Earth since. Somewhat concurrently free O_2 came into the atmosphere and oceans as widespread green-plant photosynthesis, primarily from microbial mats and stromatolites, began as far back as 3500 Ma (Awramik, 1992). Complex biogeochemical processes within the sediments and the air and waters of the Earth, involving the carbon, sulfur, and other cycles were underway. These cycles are now beginning to be understood (Walker, 1977; Schopf, 1983; Holland, 1984; Gregor et al., 1988; Schopf and Klein, 1992; Sundquist, 1993). The latter part of the Archean and the earlier part of the Paleoproterozoic was indeed a time of transition as shown by these studies and those of banded iron formations (BIFs) and uraninite deposits. Although the transition from the presumed disorganized tectonic styles prevailing before continents were assembled into styles resembling Phanerozoic plate tectonics probably took place over far more than 500 m.y., the date of 2500 Ma has been selected by international commissions (Plumb, 1991) for the boundary between the Archean and the Proterozoic as a convenience overlooking the likelihood and duration of a long transition.

Within Paleoproterozoic time Walker et al. (1983) argue for a step increase in O_2 concentration with respect to the present atmospheric level from much less than 10^{-3} O_2 to about 10^{-2} beginning at ca. 2300 Ma, a time when glaciers are viewed here as waxing at places. Kasting (1993) advocates a more gradual transition from much less than 10^{-3} O_2 concentration at ca. 2400 Ma to one about 10^{-2} by 1700 Ma. In addition, the ozone layer may well have developed as this transition progressed, shielding evolving life forms from injurious amounts of ultraviolet radiation and therefore speeding up both the growth of photosynthesizing life and the amount of O_2 (Schopf, 1992; Kasting, 1993). The Paleoproterozoic glacial intervals appear to have come along within this long transition so questions arise: To what degree are the glaciations the consequence of the rise of O_2 in the atmosphere and the commensurate reduction in CO_2 or the result of associated happenings, such as geotectonic or biogeochemical events?

Here again the geologic record is too fragmentary. The older Paleoproterozoic Huronian and Karelian glaciations are considered to have started at continental margins (Marmo and Ojakangas, 1984; Young, 1988, 1991) and glaciers seem to have spread widely upon subdued continental margins as iciness culminated and then waned. They may have begun to grow earlier upon marginal mountain ranges and uplands but any record of them has long disappeared.

This long transition from an Archean world to a Proterozoic

world, termed the "Proterozoic Revolution" by Cloud (1988), lasted for more than 500 m.y., from at least near 3000 Ma to after 2500 Ma. The revolution, however, was probably episodic and may have lasted much longer and may well have included the opening and closing of seas and the rising and wearing away of mountain ranges, and perhaps the evidence of older glacial episodes (Lowe et al., 1992). Cratonization in southern Africa seems to have started with the Kaapvaal craton near 3700 Ma (de Wit et al., 1992). The kernel of the Canadian Shield formed somewhat later, and its initial cratonization ended at ca. 2500 Ma (Goodwin, 1991). Continental glaciers need a continent to sit upon and so the first glaciations can have occurred only when appropriate sites upon them began to gather snow. The earliest known record of presumed glacial ice, that of the Witwatersrand-Pongola sequences of southern Africa, began early during the revolution nearly 3000 m.y. ago. When parts of early continents stood relatively high and open seas were at hand to provide sources for evaporation, the stage was set for the first glaciation.

CONCLUSIONS: CAUSES OF CLIMATE VARIATIONS

Ice was present on Earth for much of its history.
Frakes and Francis, 1988, p. 547

*There is no single cause . . . that controls
the timing of glaciations in Earth history.*
Eyles and Young, 1994, p. 21–22

The Earth evolved through a series of chance events . . . .
Taylor, 1998, p. 186

For the past 3 b.y. of Earth history, there is a piecemeal and fragmentary record of icy climates broken by long times of relative warmth. This record and the inference that plate-tectonic activity has prevailed indicate that glaciers have been somewhere on mountain ranges whenever and wherever moisture-laden air flowed up upon them causing lasting snow. Only at irregular intervals have ice tongues from glaciers expanded to reach the sea. Glaciers upon Earth are therefore the norm and not the exception.

Many interrelated processes have controlled climate and, in turn, glaciation. These are mainly rooted in the tectonic mobility of our planet's crust acting in concert with biogeochemical changes resulting from the evolution of life, but interspersed now and then with unusual events such as bolide impacts and huge volcanic outpourings. Review of the record indicates that climate has remained within bounds between warmer and cooler episodes consisting of long intervals of slow change broken by shorter intervals of rapid change. In short, the air-ocean-land-life system has not been forced for long from a state of quasi-equilibrium, a system of near-homeostasis. Throughout this 3 b.y., the interplay of the many processes influencing climate lay in near-balance within the critical borderland between stasis on the one hand and near-chaos on the other (Shaw, 1987; Cohen and Stewart, 1994; Lovelock and Kump, 1994; Kauffman, 1995; Bak, 1996). Perhaps the sequence of events has largely come about by chance (Taylor,

1998). At no time has the system been so disturbed, for example, by bolide impacts, that it has been unable to restore itself to a status similar to the norm. A markedly different climate state has not resulted. At times, however, if two or more strong forcing factors—of the many possible—operated together, the climate changed toward the extremes of variability. Perhaps at these unusual times notable biological extinctions were the result.

The record of pre-Mesozoic ice ages comes primarily from distinctive facies laid down in basinal strata and from geochemical proxies and indicators. Only rarely are glacially striated pavements and glacially scoured landforms preserved to fix the position of glaciers themselves. In contrast, identifiable glacial debris, caught up in icebergs and carried in ocean currents for many tens or even hundreds of kilometers and then dumped into deep basinal strata, provides evidence of ancient glaciation. This evidence discloses that glaciers thrived somewhere upcurrent but not necessarily upon the nearest land. Information is needed about the flow patterns of ocean currents, information that is seldom available.

Tectonic movements rooted in the internal thermodynamics of the Earth constitute the main cause of climate change. These determine the positioning of continents and their shapes and the bathymetry of the ocean floor, and the flux of carbon dioxide and other gases from the Earth's mantle. Wherever and whenever there was high ground upon continents at high latitudes, ice accumulated. Although the land record of these upland glaciers has largely been eroded, a sparse record comes down to us in strata laid down in basins both near and far from the glaciers themselves. To be useful to us, the record preserved in these deep basins must have been sheltered through time from destruction by uplift and erosion, or not destroyed by very deep burial and metamorphism. The record then must be uplifted and exposed so that we can study it. For very ancient times, in view of the unlikelihood of this sequence of events, we are indeed fortunate to have at hand today any record at all.

For no ice age is it possible to assign a predominating single cause. Instead, a combination or short list of concurrent events seems primarily responsible for some, but for others the interplay of many factors is required. Here it is appropriate to review the several circumstances that bring about the swings of climate toward coolness. The arrangement of continents upon the Earth is of paramount importance. Continental masses or cratons in high latitudes, especially if they have mountains upon them, make ideal sites for ice accumulation. If cratons are arranged so that ocean currents are forced to carry tropical waters into polar regions, the fundamental distribution of heat throughout the system is affected. Warm water carried to polar seas evaporates easily and makes a ready source for moisture to drift ashore and precipitate as snow that can then consolidate into ice. Water vapor is the most important greenhouse gas, and these arrangements largely control its distribution. In addition, they also control or influence the location of belts of high and low albedo, which in turn influence the heat input to the Earth system. During Paleozoic ice ages, for example, the supercontinent of Gondwana drifted across the South Pole so that there were ample sites for continental ice, especially upon rift coasts bordering open seas.

The shapes of ocean bottoms and the topography of continents are significant in distributing heat within the system. They control the regional climate, as in the North Atlantic region today, which then influences the generation of oceanic bottom waters, regional cloudiness, and the poleward flow of surface waters and the equatorward flow of bottom waters. This oceanic conveyor-belt system plays a dominating role in planetwide heat distribution, and bathymetric highs and lows determine the flow arrangements. Without an east-west bathymetric barrier upon the ocean floor, cold polar water may reach equatorward and farther. Coasts with a near-longitudinal orientation, and located so that winds blow obliquely toward them, may bring about upwelling of deep waters. This situation prevails today along the western coasts of southern Africa, southern South America, and North America. Monsoons result from the arrangement of continents with respect to ocean and air flow, as occurs today where the continental mass of India lies next to the broad and warm Indian Ocean with the Himalaya Mountains and Tibetan Plateau bordering to the north. This arrangement in southern Asia and the Indian Ocean is deemed as significant in bringing about the Late Cenozoic Ice Age (Raymo and Ruddiman, 1992; Raymo, 1994b). How often have similar arrangements occurred in remote geologic times? In short, the arrangement of continental cratons upon Earth, the location of pathways for ocean currents between them, and the shape of bathymetry and topography are significant in setting the stage for heat exchange and climate cool episodes, including glaciations. These arrangements in turn determine the location of seas for evaporation and lands for snow accumulation. They control the distribution of water vapor, the most important greenhouse gas, by means of influencing cloudiness and albedo, precipitation belts, and the shape of the conveyor-belt system.

The premise is defensible that the tectonic processes that cause the positioning of land masses or the uplift of mountain ranges to guide ocean and air circulation have remained about the same during the 3 b.y. of climate history reviewed here. Only the paces and patterns and intervals between vigorous tectonic and volcanic activities have changed, but not the basic processes themselves. Continents have assembled and dispersed. Cratons have moved about and have occupied polar regions at times, and then have drifted elsewhere. This tectonic mobility is viewed as similar to the plate-tectonic activity of today, and characterizes the movement of crustal units ever since continents were first formed and amalgamated into cratons.

The record suggests, however, that there were long periods of relative stasis with but slow changes in tectonism and climate. These quiet times were punctuated by shorter episodes of marked change, including plate-tectonic reorganizations such as when Gondwana merged with northern cratons to form Pangea. Episodes of voluminous volcanism may have accompanied such changes. Concurrently, biogenic processes modified the composition of the atmosphere and ocean and surficial sediments. On the other hand, occasional bolide impacts, perhaps interspersed irregularly or under the control of an astronomical cyclicity, wreaked their damage from time to time. It is also likely that orbital arrangements have played their role in affecting the solar influx to the Earth during the 3000 m.y. under treatment here. For pre-Phanerozoic times, however, there is as yet no convincing record within dated and timed stratal sequences to demonstrate incontrovertibly Milankovitch-type variations. Nor does the obscure and piecemeal record as now known require a tipped spin axis to explain low-latitude glaciations in pre-Phanerozoic times although continuing studies may enliven the hypothesis (G. E. Williams, 1975b, 1993, 1994; D. M. Williams et al., 1998).

Since the revolutionary discoveries of Hutton, Lyell, and Darwin during past centuries, the doctrine of uniformitarianism has guided geologic thought and research paths. This is the concept that Earth processes we observe operating today can account for all geological features we find about us, provided there is enough geologic time. Today we accept that these processes are interrupted now and then by bolide impacts or extreme combinations of terrestrial events, such as violent ocean-water overturns. A doctrine of "interrupted uniformitarianism" is now more appropriate. Although geologists depend upon the basic premise that the products we observe or infer today formed by the same processes in the geologic past, we cannot be certain that we have identified all of the processes and products. After all, it has only been during the past two decades that the significance of bolide impacts, orbital variations, and conveyor-belt circulations in the oceans has been satisfactorily demonstrated (Hays et al., 1976; Alvarez et al., 1980; Silver and Schultz, 1982; Broecker and Denton, 1989). What new processes or products, such as the discovery of new revealing isotopic proxies or geochemical combinations or ratios, may be found during the decades ahead? The Earth system is extremely complex with many interacting influences and simple explanations are unlikely. Much of the scientific search for explanations of past geologic history has followed the human urge to find simple explanations of the way nature works, an urge that is now being subdued and even discarded (e.g., Bak, 1996). Explanations of past climates and other ancient events lie now in understanding better the factors contributing to complexity and to feedbacks leading to earthly homeostasis. The question arises: To what extent has the chance combination of several factors combined to cause ice ages (Taylor, 1998)?

As we go back into remote time—and into the Archean—only rarely can geologists establish synchroneity of geologic events over broad regions. We are just beginning to appreciate the meaning and implications of the profound abyss of time, and the scarcity of records of events during the vast depths of time as we delve ancientward. Only rarely are events dated satisfactorily, and most climatological reconstructions of the very ancient past encompass tens of millions of years and are far from synoptic. Reconstructions are now only constrained in time when there are widespread and datable ash falls in the stratal record, or identifiable rapid changes in sea level, or sharp isotopic or paleomagnetic excursions that allow interregional correlations. For any geological interval depicted in a paleogeographic reconstruction with a timing uncertainty of tens of millions of years, such as most advanced tentatively for pre-Phanerozoic intervals, variations in the air-ocean-land system are likely to have been quite considerable during the interval. In view of the absence of a record of near-synchronous events, the wide uncertainties of correlation pose serious difficulties to Earth historians. Correlations closer than 10 m.y. or 20 m.y. are rare, and yet, following our premise that tectonic and other geologic activities of the remote past were similar to those operating today and operated about as rapidly, our scraps of data probably do not apply to events that were indeed contemporaneous. Moreover, quick-acting events that are now recognized in the late Cenozoic record, such as the sudden collapse of ice sheets or Heinrich events, are not yet identified in the pre-Mesozoic glacial record. Telltale records of sudden events such as bolide impacts and ocean-water overturns are just now coming to light.

The relative abundances of the greenhouse gases, mainly water vapor, carbon dioxide, and methane, have varied through time and contribute significantly to climate change, as pointed out in previous chapters. These gases have probably increased and decreased in relative abundance within bio-friendly limits since our remote cutoff date here at ca. 3000 Ma. During the long mid-Proterozoic nonglacial gap, carbon dioxide may have been more abundant than usual and the Earth may have witnessed an equitable climate. Later, within the Neoproterozoic, climate entered a time of marked fluctuations. During the Phanerozoic, glaciation flourished at times when its proportion was low.

The freeboard of the continents and the relative rise and fall of sea level are also influential. During the late Paleozoic, Gondwana stood topographically high in a polar position, and perhaps sited atop a mantle superswell so that sea level was low (Veevers and Powell, 1987). Such circumstances may have occurred in pre-Phanerozoic times as well and in part may be disclosed by variations in isotopic ratios, such as those of strontium, that reveal information on weathering and erosion rates of uplifted cratons. Sea-level variations are also influential in guiding the carbon cycle: In Phanerozoic time, after the Silurian and the burgeoning of land plants, carbon was sequestered on vegetated coastal near-flatlands as sea level rose and then exposed for redistribution into the sea as sea level dropped and erosion ensued. Variations of other isotopic proxies as well tell us much concerning the behavior of the complex tectonobiogeochemical system. In addition to strontium and carbon cycles, the sulfur, oxygen, and phosphorous cycles affect fluids of the Earth and respond to basic changes in the ocean-air-earth system and also contribute to them.

Changes resulting from the evolution of life and its biogeochemical processes and products are especially important in affecting climate. Oxygen, first in chemical combination with other elements, and then as a free gas, came into the atmosphere and oceans during the long durations of Archean and Proterozoic time (Cloud, 1988). Toward the end of the Proterozoic, oxygen was significantly involved in the metabolism of animals. It has accumulated irregularly through Phanerozoic time, and today free oxygen makes up about a fifth of the air. The overturn of deep anoxic ocean waters during the transition of the Ordovician into the Silurian and concomitant release of organic carbon both may have contributed to extinction of many sea creatures and to iciness as the carbon dioxide proportion was decreased. This overturn of deep, anoxic ocean waters was probably largely the result of unusual bathymetric and paleogeographic arrangements. During times in the late Neoproterozoic, following worldwide glaciation, the rapid buildup of carbon dioxide is suggested as responsible first for warming, then melting of ice, and then deposition of tillite-capping carbonates (Hoffman et al., 1998a,b).

Volcanic eruptions have occurred throughout geologic time, but during some intervals, as near the end of the Permian period, they were especially voluminous. At times island and continental volcanic arcs were located near atmospheric convergences, such as those at the subtropical convergence, so that their gases and dusts were carried high into the troposphere to spread widely and affect the heat flux from the sun reaching the Earth's surface. The albedo from the tops of high clouds and dust layers increased even though at the same time greenhouse gases increased in abundance beneath them. Albedo also increased when large continental areas were located in desert belts and when cloudy belts flourished across the oceans. The Saharan region today, along with deserts in western North America and Asia, possesses a bright reflectivity and therefore reduces the heat input into the system. In addition, when continental expanses are covered with snow, or when sea ice is abundant, the albedo is increased. The polar regions at present have small areas relative to temperate and equatorial belts and do not affect the total Earth albedo as much as low-latitude belts. Moreover, polar regions receive only a glancing component of heat radiation from the sun because the rotational axis of the Earth makes a high angle with the plane of the ecliptic. These factors also contribute to the strength or weakness of seasonal change: the seasonality (Crowley et al., 1986, 1991; Crowley and Baum, 1991a). Discussion prevails on whether at times during the pre-Phanerozoic a tipped spin axis may have reversed this arrangement (G. E. Williams, 1975b, 1993, 1994; D. M. Williams et al., 1998), or whether rapid slippage of the Earth's outer shells took place as inferred from investigations of true polar wander (Kirschvink et al., 1997).

Many such factors are now being considered in computer modeling (e.g., Crowley and North, 1991; Ganopolski et al., 1998). Before the current decade modeling had been based on unacceptable simplifications, such as those that do not consider the changes in ocean-surface temperatures as ocean currents have distributed heat within the air-ocean-land system. Model usefulness is dependent upon faithfulness in reproducing the flow of air and ocean currents, which in turn is based upon the reliability of the input map showing the distribution of land and sea and their topography and bathymetry. For times of ancient ice ages, data are just now becoming available for geoscientists to have confidence in maps that are basic to model input. The time is now upon us when computer modeling will significantly guide geoscientists in their search for understanding of ancient climates. We hope models will soon be able to consider realistic land and sea arrangements, oceanic bathymetery, belts of high and low albedo, the location of mountain ranges and volcanoes, regions of upwelling, and other significant features in running a model program. Geologists need guidance from modelers on what to look for in the historical record.

There is little reason to advocate long-term climate cycles with similar durations back into very ancient geologic time although orbital cycles seem established at least into the early Mesozoic. As shown on Figure l, the icy intervals occurred in the following irregular sequence working backward in time: Between the present and back to ca. 43 Ma; from ca. 105 Ma to ca. 140 Ma (with an approximate duration of 35 m.y.); ca. 160–175 Ma (ca. 15 m.y. duration); ca. 188–195 Ma (ca. 7 m.y.); ca. 256–338 Ma (ca. 82 m.y.); ca. 353–363 Ma (ca. 10 m.y.); ca. 429–445 Ma (ca. 16 m.y.); ca. 520–950 Ma (several during this ca. 430 m.y. interval); (a possible nonglacial interval of ca. 1250 m.y.); ca. 2200–2400 Ma (ca. 200 m.y.); and ca. 2914–2990 Ma (ca. 75 m.y. or ca. 76 m.y.). Some of these glaciations are stronger and spread farther geographically than others and include recorded waxings and wanings within the glacial intervals. Some ancient glaciations are recorded perhaps only because of the fortunes of preservation and later uplift and exposure. Many are not yet dated adequately. The intervals between these recognized icy times, again going back in time, are: ca. 62, ca. 61, ca. 15, ca. 66, ca. 75, several during ca. 430 m.y. interval, ca. 1250 m.y., and ca. 514 m.y. A cyclic regularity in these dates is not discernible so it seems inappropriate to advocate time cycles or supercycles for ice ages, or in turn, for major episodes of climate change or tectonic phases as suggested by some geoscientists (e.g., Fischer, 1982, 1986; Rampino and Stothers, 1984; Nance et al., 1986; Worsley et al., 1986; Veevers, 1990; Worsley and Kidder, 1991; Socci 1992). The recurrence intervals of ice ages do not lend support to the concept of periodicity in geologic history whether caused by terrestrial or extraterrestrial events (cf., Shaw, 1987). To stay with descriptive nomenclature and to avoid genetic implications, I prefer reference to icy or cool times and to warm times or intervals, rather than to "greenhouse" times, for example, and with no implications concerning dependence on greenhouse gases although their relative proportion is indeed important.

I therefore prefer an explanation for climate change without megacycles, and one characterized by irregular repetitions. These noncyclic repetitions are viewed as rooted primarily in noncyclic deep thermodynamic overturns and events within the Earth's interior. Commotions in mantle motions motivated the mobility of the lithosphere when the Earth's internal heat engine moved from moderate movements to rapid movements and back again from time to time. Plate rearrangements, supercontinent fragmentation, long-term risings or lowerings of sea level, and other major events took place whenever mantle circulations crossed some critical thermodynamic boundary. The convective convulsions responsible for these major events resulted in geographic and bathymetric rearrangements that in turn influenced the circulation of heat, primarily the heat that came from the sun but also that from the Earth's interior. In the meantime, during this long

history, life's evolution followed along its irregular path through time and fed back chemical products into the system. Superposed upon these long-term repetitions are shorter cycles such as those resulting from orbital variations and perhaps other extraterrestrial influences. From time to time bolides crashed into the Earth and there may have been intervals when true polar wander contributed to modified surface environments. Geologists are now challenged to find ways to recognize and document the record of these many influences and to separate them. To what extent has the chance coincidence or superposition of different events or processes—both those rooted in the internal geodynamics of the Earth and those rooted in extraterrestrial happenings—brought about severe climate changes?

In short, ice ages and climate changes upon our planet throughout the past 3 b.y. are the result of interplays between terrestrial and extraterrestrial influences, all operating concomitantly with life's evolution and its associated biogeochemical changes. The many interplaying processes have gone on with different styles and paces, now fast and now slowly, and interrupted irregularly by times of especially rapid change. *Tectonobiogeochemical processes operating irregularly throughout geologic time and moderated by orbital variations and interrupted occasionally by bolide impacts are the basic causes of climate change.* Complexity and variability are the hallmarks of these processes, but operating within bounds. Nonetheless, all of the interplaying events and processes functioning through this long reach of geologic time have remained within limits, never too warm and never too cool, so life could flourish. Variations in climate on into the future, including ice ages, will remain within these bounds. Human industrial, agricultural, and deforestation activity will only slightly perturbate the climate system, and only temporarily as humankind is forced to give up its dependence on hydrocarbons and coal for its energy. Within the coming century our influence on the carbon system will probably come to an end and the buffering processes of weathering and the hydrologic cycle that have prevailed during geologic time will once again take over. Although humanity will be inconvenienced, the basic health of planet Earth will continue.

ACKNOWLEDGMENTS

I have been privileged to have lived during a time when there have been widespread appreciation and support of an academic system emphasizing a combination of research and teaching. Extensive travel to examine strata critical to understanding icy parts of the geological record has therefore been encouraged. The policies of the University of California and the U.S. National Science Foundation, in particular, have permitted and partly supported such wanderings. Sabbatical leaves, funds to attend scientific meetings, and excellent libraries and facilities are among the benefits I have enjoyed as a member of the faculties of the University of California at both Los Angeles and Santa Barbara over the past 52 years. A Fulbright Award to Austria (1953–1954), a Guggenheim Fellowship to Switzerland (1954), a National Science Foundation Senior Postdoctoral Fellowship to Scotland (1960), and visiting lectureships or research projects to Bolivia, Brazil, China, Colombia, many places in Europe, India, Peru, New Zealand (especially an Erskine Fellowship to the University of Canterbury in 1991), Trinidad, Venezuela, and elsewhere have provided me with opportunities to spend days at each venue on nearby outcrops. Research grants from the U.S. National Science Foundation have supported work by me and my students in North America and upon the Gondwanan continents, including Antarctica. Field trips under the leadership of local experts, whom I have met while lecturing for continuing education programs of the American Association of Petroleum Geologists, the Geological Society of South Africa (1982 Du Toit Memorial Lecturer), the International Association of Sedimentologists (1993–1994 Special Lecturer to venues in Europe, New Zealand, Australia, and South Africa), and several petroleum companies, have been most helpful. Participation in International Geological Correlation Project 38 (Pre-Pleistocene Tillites, W. B. Harland, Chm.), Project 118 (Upper Precambrian correlations, R. Trompette, Chm.), and Project 260 (Earth's Glacial Record, M. Deynoux, Chm.) has been especially rewarding. A trip to China in 1981 under the auspices of the U.S. Committee on Scholarly Communication with the People's Republic of China, administered by the National Academy of Sciences and arranged by the Chinese Academy of Geological Sciences (Li Tingdong, President) allowed me to study many Sinian stratigraphic sections accompanied by experts, including Professor Xing Yusheng.

I have very much profited from many field excursions arranged by geological organizations here and there, and especially from extensive personal trips guided by Arturo Amos (Argentina), Victor Gostin and Ken Plumb (Australia), Romiro Suárez-Soruco, Carlos Oviedo Gómez, and Carlos Vargas Flores (Bolivia), Paulo Figueiredo and Tony Rocha-Campos (Brazil), Naseer Ahmad, Hal Borns, and Brad Hall (India), Johan Petter Nystuen (Norway), Marcos Fernandez-Dávila and Hugo Valdavia (Peru), Doug Rankin (Virginia, USA), Lawrie Minter, Tom Stratten, Johan Visser, and Victor Von Brunn (South Africa), and Tony Spencer (Scotland). Former students and postdoctoral researchers who have undertaken studies with me bearing on the Earth's glacial history include Mário Vicente Caputo, Nicholas Christie-Blick, Donald A. Coates, James C. Dawson, Lawrence A. Frakes, Paul Karl Link, Julia M. G. Miller, and A. Thomas Ovenshine. To the many friendly and helpful colleagues and organizations I extend my sincere thanks. Drafts of parts or the whole of this paper have very much benefited from the critical comments of Mário Caputo, Max Deynoux, Nick Eyles, Larry Frakes, Paul Hoffman, Paul Link, Julia Miller, Henry Schwarcz, Johan Visser, Victor Van Brunn, Grant Young, and an anonymous reviewer. Elizabeth O'Black Gans did the computer drafting and I took all of the photographs.

UCSB Institute for Crustal Studies Publication 0264-71TC.

REFERENCES CITED

Aalto, K. R., 1971, Glacial marine sedimentation and stratigraphy of the Toby Conglomerate (upper Proterozoic) southeastern British Columbia, northwestern Idaho, and northeastern Washington: Canadian Journal of Earth Sciences, v. 8, p. 753–787.

Abed, A. M., Makhloug, I. M., Amireh, B. S., and Khalil, B., 1993, Upper Ordovician glacial deposits in southern Jordan: Episodes, v. 16, p. 316–328.

Adie, R. J., 1952, The position of the Falkland Islands in a reconstruction of Gondwanaland: Geological Magazine, v. 89, p. 401–410.

Agassiz, L., 1840, Études sur les glaciers: Neuchâtel, Switzerland, privately published, 346 p.

AGU, 1992, American Geophysical Union Special Report: Volcanism and climate change. Report of Chapman Conference on climate, volcanism, and global change, under chairmanship of Stephen Self: Washington, D.C., American Geophysical Union, 27 p.

Aitken, J. D., 1991a, Two Late Proterozoic glaciations, Mackenzie Mountains, northwestern Canada: Geology, v. 19, p. 445–448.

Aitken, J. D., 1991b, The Ice Brook Formation and post-Rapitan, Late Proterozoic glaciation, Mackenzie Mountains, Northwest Territories: Geological Survey of Canada Bulletin 404, 43 p.

Aleinikoff, J. N., Zartman, R. E., Rankin, D. W., Lyttle, P. T., Burton, W. C., and McDowell, R. C., 1991, New U-Pb zircon ages for rhyolite of the Catoctin and Mount Rogers formations: more evidence for two pulses of Iapetus rifting in the central and southern Appalachians: Geological Society of America Abstracts with Programs, v. 23, no. 1, p. 2.

Aleinikoff, J. N., Horton J. W., Jr., and Walter, M., 1996, Middle Proterozoic age for the Montpelier anorthosite, Goochland terrane, eastern Piedmont, Virginia: Geological Society of America Bulletin, v. 108, p. 1481–1491.

Algeo, T. J., Berner, R. A., Maynard, J. B., and Scheckler, S. E., 1995, Late Devonian oceanic anoxic events and biotic crises: "rooted" in the evolution of vascular land plants? GSA Today, v. 5, p. 64–66.

Al-Laboun, A. A., 1987, Unayzah Formation: A new Permian-Carboniferous unit in Saudi Arabia: American Association of Petroleum Geologists Bulletin, v. 71, p. 29–38.

Almeida, F. F. M. de, and Mantovani, M. S. M., 1975, Geologia e geocronologia do Granito São Vicente, Mato Grosso: Anais da Academia Brasileira de Ciências, Rio de Janeiro, v. 47, p. 451–458.

Altermann, W., 1986, The Upper Paleozoic pebbly mudstone facies of peninsular Thailand and western Malaysia—Continental margin deposits of Palaeoeurasia: Geologische Rundschau, v. 76, p. 945–948.

Alvarenga, C. J. S. de, and Trompette, R., 1992, Glacially influenced sedimentation in the Later Proterozoic of the Paraguay belt (Mato Grosso, Brazil): Palaeogeography, Palaeoclimatology, Palaeoecology, v. 92, p. 85–105.

Alvarez, L. W., Alvarez, W., Asaro, F., and Michel, H. V., 1980, Extraterrestrial cause of the Cretaceous-Tertiary extinction: Science, v. 208, p. 1095–1108.

Alvarez, P., and Maurin, J.-C., 1991, Sedimentation and tectonics in the Upper Proterozoic basin of Comba (Congo): sequence stratigraphy of the West Congolian Supergroup and strike-slip damping model related to the Pan-African orogenesis: Precambrian Research, v. 50, p. 137–171.

Alvarez, W., 1997, *T. rex* and the crater of doom: Princeton, New Jersey, Princeton University Press, 185 p.

Alvarez, W., Claeys, P., and Kieffer, S. W., 1995, Emplacement of Cretaceous-Tertiary boundary shocked quartz from Chicxulub crater: Science, v. 269, p. 930–935.

Amos, A. J. and López Gamundi, O., 1981, Late Paleozoic Sauce Grande Formation of eastern Argentina, *in* Hambrey M. J., and Harland, W. B., eds., Earth's pre-Pleistocene glacial record: Cambridge, United Kingdom, Cambridge University Press, p. 872–877.

Anandakrishman, S., Blankenship, D. D., Alley, R. B., and Stoffa, P. L., 1998, Influence of subglacial geology on the position of a West Antarctic ice stream from seismic observations: Nature, v. 394, p. 62–65.

Anderson, D. L., 1993, Helium-3 from the mantle: Primordial signal or cosmic dust: Science, v. 261, p. 170–176.

Anderson, D. L., 1994, Superplumes or supercontinents?: Geology, v. 22, p. 39–42.

Anderson, J. B., 1983, Ancient glacial-marine deposits: their spatial and temporal distribution, *in* Molnia, B. F., ed., Glacial-marine sedimentation: New York, Plenum Press, p. 3–92.

Anderson, J. B., and Ashley, G. M., eds., 1991, Glacial marine sedimentation: paleoclimate significance: Geological Society of America Special Paper 261, 232 p.

Anderson, M. M., 1972, A possible time span for the Late Precambrian of the Avalon Peninsula, southeastern Newfoundland in the light of worldwide correlation of fossils, tillites, and rock units within the succession: Canadian Journal of Earth Sciences, v. 9, p. 1710–1726.

Anderson, S. P., 1997, Chemical weathering in glacial environments: Geology, v. 25, p. 399–402.

Anderton, R., 1982, Dalradian deposition and the late Precambrian-Cambrian history of the North Atlantic region: A review of the early evolution of the Iapetus Ocean: Geological Society of London Journal, v. 139, p. 421–431.

Andreae, M. O., 1996, Raising dust in the greenhouse: Nature, v. 380, p. 389–390.

Andrews, J. T., ed., 1974, Glacial isostasy: Stroudsburg, Pennsylvania, Dowden, Hutchinson and Ross, Inc., 491 p.

Andrews, J. T., 1990, Fiord to deep sea sediment transfers along the northeastern Canadian continental margin: models and data: Geographie Physique et Quaternaire, v. 44, p. 55–70.

Andrews, J. T., and Matsch, C. L., 1983, Glacial marine sediments and sedimentation: An annotated bibliography: Norwich, England, Geo Abstracts, 227 p.

Arbey, F., and Tamain, G., 1971, Existence d'une glaciation siluro-ordovicienne en Sierra Morena (Espagne): Comptes Reondus de l'Académie des Sciences, Paris, v. 272, p. 1721–1723.

Armin, R. A., and Mayer, L., 1993, Subsidence analysis of the Cordilleran miogeocline: Implications for timing of late Proterozoic rifting and amount of extension: Geology, v. 11, p. 702–705.

Armstrong, R. L., 1971, Glacial erosion and the variable isotopic composition of strontium in sea water: Nature, Physical Sciences., v. 230, p. 132–133.

Arnold, F., Burke, T., and Qui, S., 1990, Evidence for stratospheric ozone-depleting heterogeneous chemistry on volcanic aerosols from El Chichon: Nature, v. 348, p. 49–50.

Arthur, M. A., 1982, The carbon cycle—controls on atmospheric CO_2 and climate in the geologic past, *in* Climate in Earth History, Studies in Geophysics: Washington, D.C., National Academy Press, p. 55–67.

Arthur, M. A., and Jenkyns, H. C., 1981, Phosphorites and paleoceanography, *in* Publications of the 26th International Geological Congress, Paris, Colloquium C-4, Geology of Oceans: Oceanologica Acta, No. SP, p. 83–96.

Asmerom, Y., Jacobsen, S. B., Knoll, A. H., Butterfield, N. J., Swett, K., 1991, Strontium isotopic variations of Neoproterozoic seawater: Implications for crustal evolution: Geochimica et Cosmochimica Acta, v. 55, p. 2883–2894.

Astini, R. A., Benedetto, J. L., and Vaccari, N. E., 1995, The early Paleozoic evolution of the Argentine Precordillera as a Laurentian rifted, drifted, and collided terrane: A geodynamic model: Geological Society of America Bulletin, v. 107, p. 253–273.

Awramik, S. W., 1992, The oldest records of photosynthesis: Photosynthesis Research, v. 33, p. 75–89.

Awramik, S. W., and Vanyo, J. P., 1986, Heliotropism in modern stromatolites: Science, v. 231, p. 1279–1281.

Axelrod, D. I., 1981, Role of volcanism in climate and evolution: Geological Society of America Special Paper 185, 59 p.

Bachtadse, V., and Briden, J. C., 1990, Palaeomagnetic constraints on the position of Gondwana during Ordovician to Devonian times, *in* McKerrow, W. S., and Scotese, C. R., eds., Palaeozoic Palaeogeography and Biogeography: Geological Society of London Memoir 12, p. 43–48.

Badenhorst, F. P., 1988, The lithostratigraphy of the Chuos mixtite in part of the southern central zone of the Damara orogen, South West Africa: Communications of Geological Survey of S.W. Africa/Namibia, v. 4, p. 103–110.

Bak, P., 1996, How nature works: New York, Springer-Verlag, 212 p.

Baker, M. B., 1997, Cloud microphysics and climate: Science, v. 276, p. 1072–1078.

Barnola, J. M., Raynaud, D., Korotkevich, Y. S., and Lorius, C., 1987, Vostok ice core provides 160,000-year record of atmospheric CO_2: Nature, v. 329, p. 408–414.

Barrett, P. J., 1991, The Devonian to Triassic Beacon Supergroup of the Transantarctic Mountains and correlatives in other parts of Antarctica, *in* Tingey, R. J., ed., The Geology of Antarctica: Oxford, United Kingdom, Clarendon Press, p. 120–152.

Barron, E. J., and Peterson, W. H., 1989, Model simulation of the Cretaceous ocean circulation: Science, v. 244, p. 684–686.

Barron, E. J., and Washington, W. M., 1984, The role of geographic variables in explaining paleoclimates: Results from Cretaceous climate model sensitivity studies: Journal of Geophysical Research, v. 89, p. 1267–1279.

Barton, E. S., Altermann, W., Williams, I. S., and Smith, C. B., 1994, U-Pb zircon age for a tuff in the Campbell Group, Griqualand West sequence, South Africa: Implications for Early Proterozoic rock accumulation rates: Geology, v. 22, p. 343–346.

Beauchamp, B., 1994, Permian climate cooling in the Canadian Arctic: *in* Klein, G. D., ed., Pangea: Paleoclimate, Tectonics, and Sedimentation during Sedimentation, Zenith, and Breakup of a Supercontinent: Geological Society of America Special Paper 288, p. 229–246.

Beauchamp, B., Davies, G. R., and Henderson, C. M., 1989, Upper Paleozoic stratigraphy and basin analysis of the Sverdrup Basin, Canadian Arctic Archipelago: Ottawa, Canada, Geological Survey of Canada Paper 89-1G, p. 105–124.

Beck, R. A., Burbank, D. W., Sercombe, W. J., Olson, T. L., and Khan, A. M., 1995, Organic carbon exhumation and global warming during the early Himalayan collision: Geology, v. 23, p. 387–390.

Beckinsale, R. D., Reading, H. G., and Rex, D. C., 1976, Potassium-argon ages for basic dykes from east Finnmark: stratigraphic and structural implications: Scottish Journal of Geology, v. 12, p. 51–65.

Bell, R. E., Blankenship, D. D., Finn, C. A., Morse, D. L., Scambos, T. A., Bozena, J. M., and Hodge, S. M., 1998, Influence of subglacial geology on the onset of a West Antarctic ice stream from aerogeophysical observations: Nature, v. 394, p. 58–62.

Bennacef, A., Beuf, S., Biju-Duval, B., de Charpal, O., Gariel, O., and Rognon, P., 1971, Example of cratonic sedimentation: Lower Paleozoic of Algerian Sahara: American Association of Petroleum Geologists Bulletin, v. 55, p. 2225–2245.

Bentley, C. R., 1998, Ice on the fast track: Nature, v. 394, p. 21–22.

Berger, A., Loutre, M. F., and Laskar, J., 1992, Stability of the astronomic frequences over the Earth's history for paleoclimate studies: Science, v. 255, p. 560–566.

Berger, A. L., Imbrie J., Hays, J. D., Kukla, G. J., and Salzman, B., eds., 1984, Milankovitch and Climate: Dordrecht, Netherlands, D. Reidel, 895 p.

Berger, W. H., and Winterer, E. L., 1974, Plate stratigraphy and carbonate line, *in* Jenkyns, H. C., and Hsü, K. J., eds., Pelagic sediments: On land and under the sea: International Association of Sedimentologists Special Publication No. 1, p. 11–48.

Berger, W. H., Eddy, J. A., and Shoemaker, E. M., 1985, Effects of extraterrestrial phenomena on the evolution of complex life on Earth, *in* Milne, D., et al., eds., The evolution of complex higher organisms: Washington, D.C., U.S. Government Printing Office, NASA Special Paper 478, p. 111–143, 177–192.

Berner, R. A., 1987, Models for carbon and sulfur cycles and atmospheric oxygen: Application to Paleozoic geologic history: American Journal of Science, v. 287, p. 177–196.

Berner, R. A., 1990, Atmospheric carbon dioxide levels over Phanerozoic time: Science, v. 249, p. 1382–1386.

Berner, R. A., 1991, A model of atmospheric CO_2 over Phanerozoic time: American Journal of Science, v. 291, p. 339–376.

Berner, R. A., 1992, Palaeo-CO_2 and climate: Nature, v. 358, p. 114.

Berner, R. A., 1993, Paleozoic atmospheric CO_2: Importance of solar radiation and plant evolution: Science, v. 261, p. 68–70.

Berner, R. A., 1995, A. G. Högben and the development of the concept of the geochemical carbon cycle: American Journal of Science, v. 295, p. 491–495.

Berner, R. A., 1997, The rise of plants and their effect on weathering and atmospheric CO_2: Science, v. 276, p. 544–546.

Berner, R. A., and Caldeira, K., 1997, The need for mass balance and feedback in the geochemical carbon cycle: Geology, v. 25, p. 955–956.

Berner, R. A., and Petsch, S. T., 1998, The sulfur cycle and atmospheric oxygen: Science, v. 282, p. 1426–1427.

Berry, W. B. N., and Boucot, A. J., 1972, Correlation of South American Silurian rocks: Geological Society of America Special Paper 133, 59 p.

Berry, W. B. N., and Boucot, A. J., 1973a, Correlation of the African Silurian rocks: Geological Society of America Special Paper 147, 83 p.

Berry, W. B. N., and Boucot, A. J., 1973b, Glacio-eustatic control of the Late Ordovician–Early Silurian platform sedimentation and faunal changes: Geological Society of America Bulletin, v. 84, p. 275–284.

Bertrand-Sarfati, J., Moussine-Pouchkine, A., Amard, B., and Kaci Ahmed, A. A., 1995, First Ediacaran fauna found in western Africa and evidence for an Early Cambrian glaciation: Geology, v. 23, p. 133–136.

Bessonova, V. Y., and Chumakov, N. M., 1968, On glacial deposits in the Late Precambrian of Belorussia: Doklady Akademii Nauk SSSR, v. 178, p. 68–71.

Beuf, S., Biju-Duval, B., De Charpal, O., Rognon, P., Gariel, O., and Bennacef. A., 1971, Les grès du Paléozoïque Inférieur au Sahara—Sédimentation et discontinuités: Évolution structurale d'un craton: Institut Française du Petrole, Science et Technique du Petrole, v. 18, 464 p.

Beukes, N. J., and Cairncross, B., 1991, A lithostratigraphic-sedimentological reference profile for the Late Archaean Mozaan Group, Pongola Sequence: application to sequence stratigraphy and correlation with the Witwatersrand Supergroup: South African Journal of Geology, v. 94, p. 44–69.

Bickford, M. E., 1988, The formation of continental crust: Geological Society of America Bulletin, v. 100, p. 1375–1391.

Bickle, M. J., 1998, The need for mass balance and feedback in the geochemical carbon cycle: Comment: Geology, v. 26, p. 477–478.

Biju-Duval, B., Deynoux, M., and Rognon, P., 1981, Late Ordovician tillites of the central Sahara, in Hambrey, M. J., and Harland, W. B., eds., Earth's pre-Pleistocene glacial record: Cambridge, United Kingdom, Cambridge University Press, p. 151–152.

Binda, P. L., and Van Eden, J. G., 1972, Sedimentological evidence on the origin of the Precambrian Great Conglomerate (Kundelungu Tillite), Zambia: Palaeogeography, Palaeoclimatology, Palaeoecology, v. 12, p. 151–168.

Bjørlykke, K., 1966, Studies on the latest Precambrian and Eocambrian rocks in Norway. 1. Sedimentary petrology of the Sparagmites of the Rena District, south Norway: Norges Geologiske Undersøkelse, v. 238, p. 5–53.

Bjørlykke, K., 1967, The Eocambrian "Reusch Moraine" at Bigganjargga and the geology around Varangerfjorden, northern Norway: Norges Geologiske Undersøkelse, v. 251, p. 18–44.

Bjørlykke, K., 1973, Glacial conglomerates of Late Precambrian age from the Bunyoro Series, West Uganda: Geologische Rundschau, v. 62, p. 938–947.

Bjørlykke, K., 1985, Glaciations, preservation of the sedimentary record and sea level changes: A discussion based on the Late Precambrian and Lower Paleozoic sequence in Norway: Palaeogeography, Palaeoclimatology, Palaeoecology, v. 51, p. 197–207.

Bjørlykke, K., and Nystuen, J. P., 1981, Late Precambrian tillites of south Norway, in Hambrey, M. J., and Harland, W. B., eds., Earth's pre-Pleistocene glacial record: Cambridge, United Kingdom, Cambridge University Press, p. 624–628.

Bjørlykke, K., Bue, B., and Elverhøi, A., 1978, Quaternary sediments in the northwestern part of the Barents Sea and their relation to the underlying Mesozoic bedrock: Sedimentology, v. 25, p. 227–246.

Blanford, W. T., Blanford, H. F., and William, T., 1856, On the geological structure and relations of the Talcheer Coal Field, in the District of Cuttack: Calcutta, India, Geological Survey of India Memoir 1, Part 1, p. 33–89.

Blum, J. D., and Erel, Y., 1995, A silicate weathering mechanism linking increases in marine $^{87}Sr/^{86}Sr$ with global glaciations: Nature, v. 373, p. 415–418.

Blunier, T., Chappellez, J., Schwander, J., Dällenbach, A., Stauffer, B., Stocker, T. F., Raynaud, D., Jouzel, J., Clausen, H. B., Hammer, C. U., and Johnsen, S. J., 1998, Asynchrony of Antarctic and Greenland climate change during the last glacial period: Nature, v. 394, p. 739–743.

Boardman, D. R., II, and Heckel, P. H., 1989, Glacial-eustatic sea-level curve for early Late Pennsylvanian sequence in north-central Texas and biostratigraphic correlation with curve for midcontinent North America: Geology, v. 17, p. 802–805.

Boardman, D. R., II, and Heckel, P. H., 1992, Glacial-eustatic sea-level curve for early Late Pennsylvanian sequence in north-central Texas and biostratigraphic correlation with curve for midcontinent North America: Reply: Geology, v. 20, p. 92–94.

Bond, G., 1981a, Late Palaeozoic (Dwyka) glaciation in the middle Zambesi region, in Hambrey, M. J., and Harland, W. B., eds., Earth's pre-Pleistocene Glacial Record: Cambridge, United Kingdom, Cambridge University Press, p. 55–57.

Bond, G., 1981b, Late Paleozoic (Dwyka) glaciation in the Sabi-Limpopo region, Zimbabwe, in Hambrey, M. J., and Harland, W. B., eds., Earth's pre-Pleistocene Glacial Record: Cambridge, United Kingdom, Cambridge University Press, p. 58–60.

Bond, G. C., Nickeson, P. A., and Kominz, M. A., 1984, Breakup of the supercontinent between 625 Ma and 525 Ma: new evidence and implications for continental histories: Earth and Planetary Science Letters, v. 70, p. 325–345.

Bond, G. C., Christie-Blick, N., Kominz, M. A., and Devlin, W. J., 1985, An early Cambrian rift to post-rift transition in the Cordillera of western North America: Nature, v. 316, p. 742–745.

Bond, G. C., Heinrich, H., Broecker, W., Labeyrie, L., McManus, J., Andrews, J., Huon, S., Jantschik, J., Clasen, S., Simet, C., Tedesco, K., Klas, M., Bonani, G., and Ivy, S., 1992, Evidence for massive discharges of icebergs into the North Atlantic ocean during the last glacial period: Nature, v. 360, p. 245–249.

Borg, S. G., and DePaolo, D. J., 1994, Laurentia, Australia, and Antarctica as a Late Proterozoic supercontinent: Constraints from isotopic mapping: Geology, v. 22, p. 307–310.

Boulton, G. S., 1972, Modern arctic glaciers as depositional models for former ice sheets: Geological Society of London Journal, v. 17, p. 361–393.

Boulton, G. S., 1978, Boulder shapes and grain-size distributions of debis as indicators of transport paths through a glacier and till genesis: Sedimentology, v. 25, p. 773–799.

Boulton, G. S., 1990, Sedimentary and sea level changes during glacial cycles and their control on glacimarine facies architecture, in Dowdeswell, J. A., and Srourse, J. D., eds., Glacimarine Environments: Processes and Sediments: Geological Society of London Special Publication 53, p. 15–52.

Boulton, G. S., and Deynoux, M., 1981, Sedimentation in glacial environments and the identification of tills and tillites in ancient sedimentary sequences: Precambrian Research, v. 15, p. 397–422.

Boundy-Sanders, S. Q., 1992, Highly calcic, oddly shaped Upper Devonian microtektites from western Yukon Territory: Eos (Transactions, American Geophysical Union), Abstracts, Supplement, v. 73, p. 32B.

Bowring, S. A., and Erwin, D. H., 1998, A new look at evolutionary rates in deep time: Uniting paleontology and high-precision geochronology: GSA Today, v. 8, no. 9, p. 1–8.

Bowring, S. A., Grotzinger, J. P., Isachsen, C. E., Knoll, A. H., Pelechaty, S. M., and Kolosov, P., 1993, Calibrating rates of Early Cambrian evolution: Science, v. 261, p. 1293–1298.

Bowring, S. A., Erwin, D. H., Jin, Y. G., Martin, M. W., Davidek, K., and Wang, W., 1998, U/Pb zircon geochronology and tempo of the End-Permian mass extinction: Nature, v. 280, p. 1039–1045.

Braakman, J. H., Levell, B. K., Martin, J. H., Potter, T. L., and van Vliet, A., 1982, Late Palaeozoic Gondwana glaciation in Oman: Nature, v. 299, p. 48–50.

Brack, P., Mundil, R., Oberli, F., Meier, M., and Rieber, H., 1996, Biostratigraphic and radiometric age data question the Milankovitch characteristics of the Latemar Cycles (Southern Alps, Italy): Geology, v. 24, p. 371–375.

Brakel, A. T., and Totterdell, J. M., 1992, The Permian palaeogeography of Australia: Australian Bureau of Mineral Resources, Geology and Geophysics Record 1990/60: Paleogeography, v. 19, 126 p., many maps and charts.

Brandt, K., 1986, Glacioeustatic cycles in the Early Jurassic?: Neues Jahrbuch für Geologie und Paläontologie Monatshefte, v. 6, p. 257–274.

Brasier, M. D., and Lindsay, J. F., 1998, A billion years of environmental stability and the emergence of eukaryotes: New data from northern Australia: Geology, v. 26, p. 555–558.

Brasier, M. D., and Singh, P., 1987, Microfossils and Precambrian stratigraphy of Maldeota, Lesser Himalaya: Geological Magazine, v. 124, p. 323–345.

Brasier, M. D., Dorjnamjaa, D., and Lindsay, J. F., 1996, The Neoproterozoic to early Cambrian in southwest Mongolia: An introduction: Geological Magazine, v. 133, p. 365–399.

Bray, J. R., 1977, Pleistocene volcanism and glacial initiation: Science, v. 197,

p. 252–254.

Brenchley, P. J., and Newall, G., 1980, A facies analysis of the Upper Ordovician regressive sequences in the Oslo region, Norway—a record of glacio-eustatic change: Palaeogeography, Palaeoclimatology, Palaeoecology, v. 31, p. 1–38.

Brenchley, P. J., Marshall, J. D., Carden, G. A. F., Robertson, D. B. R., Long, D. G. F., Meidla, T., Hints, L., and Anderson, T. F., 1994, Bathymetric and isotopic evidence for a short-lived Late Ordovician glaciation in a greenhouse period: Geology, v. 22, p. 295–298.

Briffa, K. R., Jones, P. D., Schweingruber, F. H., and Osborn, T. J., 1998, Influence of volcanic eruptions on Northern Hemisphere summer temperatures over the past 600 years: Nature, v. 393, p. 450–455.

Brodzikowski, K., and van Loon, A. J., 1991, Glacigenic sediments, Developments in sedimentology, v. 49: Amsterdam, Elsevier, 674 p.

Broecker, W. S., 1989, The salinity contrast between the Atlantic and Pacific Oceans during glacial time: Paleoceanography, v. 4, p. 207–212.

Broecker, W. S., 1994, Massive iceberg discharges as triggers for global climate change: Nature, v. 372, p. 421–449.

Broecker, W. S., 1997, Will our ride in the greenhouse future be a smooth one?: GSA Today, v. 7, no. 5, p. 1–7.

Broecker, W. S., and Denton, G. H., 1989, The role of ocean-atmosphere reorganizations in glacial cycles: Geochimica et Cosmochimica Acta, v. 53, p. 2465–2501.

Broecker, W. S., Peteet, D. M., and Rind, D., 1985, Does the ocean-atmosphere system have more than one stable mode of operation?: Nature, v. 315, p. 21–26.

Brookfield, M. E., 1993, Neoproterozoic Laurentia-Australia fit: Geology, v. 21, p. 683–686.

Brouwers, E. S., Clemens, W. A., Spicer, R. A., Ager, T. A., Carter, L. D., and Sliter, W. V., 1987, Dinosaurs on the North Slope, Alaska: High latitude latest Cretaceous environments: Science, v. 237, p. 1608–1610.

Browning, J. V., Miller, K. G., and Pak, D. K., 1996, Global implications of lower to middle Eocene sequence boundaries on the New Jersey coastal plain: The icehouse cometh: Geology, v. 24, p. 639–642.

Brownlee, D. E., 1995, A driver of glaciation cycles: Nature, v. 378, p. 558.

Bryson, R. A., and Goodman, B. M., 1980, Volcanic activity and climatic changes: Science, v. 207, p. 1041–1044.

Buggisch, W., 1991, The global Frasnian-Famennian "Kellwasser Event": Geologische Rundschau, v. 80, p. 49–72.

Burbank, D. W., Derry, L. A., and France-Lanord, C., 1993, Reduced Himalayan sediment production 8 Myr ago despite an intensified monsoon: Nature, v. 364, p. 48–50.

Burchfiel, B. C., Cowan, D. S., and Davis, G. A., 1992, Tectonic overview of the Cordilleran orogen in the western United States, *in* Burchfiel, B. C., Lipman, P. W., and Zobeck, M. L., eds., The Cordilleran Orogen: Conterminous U.S.: Boulder, Colorado, Geological Society of America, The Geology of North America, v. G-3, p. 407–480.

Burdette, J. W., Grotzinger, J. P., and Arthur, M. A., 1990, Did major changes in the stable-isotope composition of Proterozoic seawater occur?: Geology, v. 18, p. 227–230.

Burke, W. H., Denison, R. E., Hetherington, E. A., Koepnick, R. B., Nelson, H. F., and Otto, J. B., 1982, Variations of seawater $^{87}Sr/^{86}Sr$ throughout Phanerozoic time: Geology, v. 10, p. 516–519.

Burton, K. W., Ling, H-F., O'Nions, R. K., 1997, Closure of the Central American Isthmus and its effect on deep-water formation in the North Atlantic: Nature, v. 386, p. 382–385.

Caby, R., and Fabre, J., 1981a, Late Proterozoic to Early Paleozoic diamictites, tillites, and associated glacigenic sediments in the Serie Pourpree of Western Hoggar, Algeria, *in* Hambrey, M. J., and Harland, W. B., eds., Earth's pre-Pleistocene glacial record: Cambridge, United Kingdom, Cambridge University Press, p. 140–145.

Caby, R., and Fabre, J., 1981b, Tillites in the latest Precambrian strata of the Touareg Shield (central Sahara), *in* Hambrey, M. J., and Harland, W. B., eds., Earth's pre-Pleistocene glacial record: Cambridge, United Kingdom, Cambridge University Press, p. 146–149.

Cahen, J., and Lepersonne, J., 1981a, Proterozoic diamictites of Lower Zäire, *in* Hambrey, M. J., and Harland, W. B., eds., Earth's pre-Pleistocene glacial record: Cambridge, United Kingdom, Cambridge University Press, p. 153–157.

Cahen, J., and Lepersonne, J., 1981b, Upper Proterozoic diamictites of Whaba (formerly Katanga) and neighbouring regions of Zambia, *in* Hambrey, M. J., and Harland, W. B., eds., Earth's pre-Pleistocene glacial record: Cambridge, United Kingdom, Cambridge University Press, p. 162–166.

Caldas V. J., 1979, Evidencias de una glaciation Precambriana en la costa sur del Peru: Segundo Congreso Geologico Chileno, Arica, Chile, Agosto 6–11, 1979, p. J29–J35.

Caldeira, K., and Berner, R. A., 1998, The need for mass balance and feedback in the geochemical carbon cycle: Reply: Geology, v. 26, p. 478.

Caldeira, K., and Kasting, J. F., 1992, Susceptibility of the early Earth to irreversible glaciation caused by carbon dioxide clouds: Nature, v. 359, p. 226–228.

Calkin, P. E. (with contributions by G. M. Young), 1995, Global glaciation chronologies and causes of glaciation, *in*, Menzies, J., ed., Past glacial environments: Sediments, forms, techniques: Oxford, United Kingdom, Butterwood-Heinemann, Ltd., p. 9–75.

Campbell, I. H., Czamanske, G. K., Fedorenko, V. A., Hill, R. I., and Stepanov, V., 1992, Synchronism of the Siberian Traps and the Permian-Triassic boundary: Science, v. 258, p. 1760–1763.

Canfield, D. E., 1998, A new model for Proterozoic ocean chemistry: Nature, v. 396, p. 450–453.

Canfield, D. E., and Teske, A., 1996, Late Proterozoic rise in atmospheric oxygen concentration inferred from phylogenetic and sulphur-isotope studies: Nature, v. 382, p. 127–132.

Caputo, M. V., 1984, Glaciação Neodevoniana no continente Gonduana ocidental: Anais do XXXIII Congresso Brasileiro de Geologia, Rio de Janeiro, v. 2, p. 725–739.

Caputo, M. V., 1985, Late Devonian glaciation in South America: Palaeogeography, Palaeoclimatology, Palaeoecology, v. 51, p. 291–317.

Caputo, M. V., 1994, Atmospheric CO_2 depletion as a glaciation and biotic extinction agent: The Devonian-Carboniferous glacial examples: IV Simpósio de geologia da Amazônia, Sociedade Basileira de Geologia, Belém, Brazil, p. 194–197.

Caputo, M. V., 1995, Anti-greenhouse effect as a glaciation and biotic crisis triggering agent: The Late Devonian and Early Carboniferous glacial cases: Resumo das Comunicações, Anais de Academia Brasileira de Ciências, v. 67, no. 3, p. 390.

Caputo, M. V., and Crowell, J. C., 1985, Migration of glacial centers across Gondwana during Paleozoic Era: Geological Society of America Bulletin, v. 96, p. 1020–1036.

Caputo, M. V., and Vasconcelos, D. N. N., 1971, Possibilidades de hidrocarbonetos no Arco Purus: Internal Report 644/A Petróleos Brasileiro S. A., Região de Exploração do Norte, Belém, Brazil (PETROBRÁS, Sistema de Informação em Exploração—SIEX 130-5164), 21 p.

Casshyap, S. M., and Kumar, A., 1987, Fluvial architecture of the upper Permian Raniganj Coal Measure in the Damoda basin, eastern India: Sedimentary Geology, v. 51, p. 181–213.

Casshyap, S. M., and Srivastava, V. K., 1987, Glacial and proglacial Talchir sedimentation in Son-Mahanadi Gondwana basin: Paleogeographic reconstruction, *in* McKenzie, G. D., ed., Gondwana Six: Stratigraphy, Sedimentology, and Paleontology: Washington, D.C., American Geophysical Union, Geophysical Monograph 41, p. 167–182.

Cerling, T. E., Quade, J., Wang, Y., and Bowman, J. R., 1989, Carbon isotopes in soils and palaeosols as ecology and palaeoecology indicators: Nature, v. 341, p. 138–139.

Chamberlin, T. C., 1899, An attempt to frame a working hypothesis of the cause of glacial periods on an atmospheric basis: Journal of Geology, v. 7, p. 545–584, 667–685, 751–787.

Charlson, R. J., Lovelock, J. E., Andreae, M. O., and Warren, S. G., 1987, Oceanic phytoplankton, atmospheric sulphur, cloud albedo and climate: Nature, v. 326, p. 655–661.

Charpentier, J. de, 1841, Essai sure les glaciers et sur le terrain erratique du bassin du Rhône: Lausanne, Switzerland, Marc Ducloux, 363 p.

Chen Jinbiao, Zhang Huimin, Xing Yusheng, and Ma Guogan, 1981, On the Upper Precambrian (Sinian Suberathem) in China: Precambrian Research, v. 15, p. 207–228.

Chen, Z., Li, Z. X., Powell, C. M., and Balme, B. E., 1993, Palaeomagnetism of the Brewer Conglomerate in central Australia, and fast movement of Gondwanaland during the Late Devonian: Geophysical Journal International, v. 115, p. 564–574.

Chorlton, W., 1982, Ice ages: Alexandria, Virginia, Time-Life Books (Planet Earth Series), 176 p.

Christie-Blick, N., 1982a, Upper Proterozoic and Lower Cambrian rocks of the Sheeprock Mountains, Utah: Regional correlation and significance: Geological Society of America Bulletin, v. 93, p. 735–750.

Christie-Blick, N., 1982b, Upper Precambrian (Eocambrian) Mineral Fork Tillite of Utah: A continental glacial and glaciomarine sequence: Discussion: Geological Society of America Bulletin, v. 93, p. 184–186.

Christie-Blick, N., 1983, Glacial-marine and subglacial sedimentation, Upper Proterozoic Mineral Fork Formation, Utah, in Molnia, B. F., ed., Glacial-marine sedimentation: New York, Plenum Press, p. 703–777.

Christie-Blick, N., and Driscoll, N. W., 1995, Sequence stratigraphy: Annual Review of Earth and Planetary Sciences, v. 23, p. 451–478.

Christie-Blick, N., Grotzinger, J. P., and von der Borch, C. C., 1988, Sequence stratigraphy in Proterozoic successions: Geology, v. 16, p. 100–104.

Christie-Blick, N., Mountain, G. S., and Miller, K. G., 1990a, Seismic stratigraphic record of sea-level change, in Sea-Level change, Studies in geophysics: Washington, D.C., National Academy Press, p. 116–140.

Christie-Blick, N., von der Borch, C. C., and DiBona, P. A., 1990b, Working hypotheses for the origin of the Woka Canyons (Neoproterozoic), South Australia: American Journal of Science, v. 290A, p. 295–332.

Christie-Blick, N., Dyson, I. A., and Von der Borch, C. C., 1995, Sequence stratigraphy and the interpretation of Neoproterozoic earth history: Precambrian Research, v. 73, p. 3–26.

Chumakov, N. M., 1968, Late Precambrian glaciation of Spitzbergen: Doklady Akademii Nauk SSSR, v. 180, p. 1446–1449.

Chumakov, N. M., 1981, Upper Proterozoic glaciogenic rocks and their stratigraphic significance: Precambrian Research, v. 15, p. 373–395.

Chumakov, N. M., 1985, Glacial events of the past and their geological significance: Palaeogeography, Palaeoclimatology, Palaeoecology, v. 51, p. 319–346.

Chumakov, N. M., 1991, Middle Siberian glacial horizon—Traces of the earliest Late Precambrian glaciation? in Report of Project 260, Earth's glacial record, International Geological Correlation Program, annual meeting and field trip, Bamako, Mali, January 7–17, 1991, Abstract, p. 115–120.

Chumakov, N. M., and Elston, D. P., 1989, The paradox of Late Proterozoic glaciations at low latitudes: Episodes, v. 12, p. 115–120.

Chumakov, N. M., and Frakes, L. A., 1997, Mode of origin of dispersed clasts in Jurassic shales, southern part of the Yena-Kolyma fold belt, North East Asia: Palaeogeography, Palaeoclimatology, Palaeoecology, v. 128, p. 77–85.

Chumakov, N. M., and Krasil'nikov, S. S., 1992, Lithology of Riphean tilloids: Urinsk uplift area, Jena region: Lithology and Mineral Resources, v. 26, p. 249–264.

Cicerone, R. J., 1994, Fires, atmospheric chemistry, and the ozone layer: Science, v. 263, p. 1243–1244.

Claeys, P., Casier, J.-G., and Margolis, S. V., 1992, Microtektites and mass extinctions: Evidence for a Late Devonian asteroid impact: Science, v. 257, p. 1102–1104.

Claoué-Long, J. C., Zichao, Z., Gougan, M., and Shaohua, D., 1991, The age of the Permian-Triassic boundary: Earth and Planetary Science Letters, v. 105, p. 182–190.

Clark, P. U., 1991, Striated clast pavements: products of deforming subglacial sediment?: Geology, v. 19, p. 530–533.

Clauer, N., and Deynoux, M., 1987, New information on the probable isotopic age of the late Proterozoic glaciation in West Africa: Precambrian Research, v. 37, p. 89–94.

Claypool, G. E., Holser, W. T., Kaplan, I. R., Sakai, M., and Zak, I., 1980, The age curves of sulfur and oxygen isotopes in marine sulfate and their mutual interpretation: Chemical Geology, v. 28, p. 199–260.

CLIMAP, 1976, The surface of the ice-age Earth: Science, v. 191, p. 1131–1144.

CLIMAP, 1981, Seasonal reconstructions of the Earth's surface at the last glacial maximum: Geological Society of America Map and Chart Series, MC-36, p. 1–18 (18 maps on 9 sheets, with text. Several scales).

Cloetingh, S., 1986, Intraplate stresses: a new tectonic mechanism for relative fluctuations of sea level: Geology, v. 14, p. 617–620.

Cloetingh, S., 1988, Intraplate stresses: a tectonic cause for third-order cycles in apparent sea level, in Wilgus, C. K., Hastings, B. S., Kendall, C. G. St. C., Posamentier, H. W., Ross, C. A., and Van Wagoner, J. C., eds., Sea-Level changes: An integrated approach: Society of Economic Paleontologists and Mineralogists Special Publication No. 42, p. 19–30.

Cloud, P., 1968, Atmospheric and hydrospheric evolution on the primitive earth: Science v. 160, p. 729–736.

Cloud, P., 1976, Major features of crustal evolution, Du Toit Memorial Lecture No. 14: Geological Society of South Africa Special Publication, v. 79 (Annexure), p. 1–32.

Cloud, P., 1988, Oasis in space: Earth history from the beginning: New York, Norton and Company, 508 p.

Coates, A. G., Jackson, J. B. C., Colins, L. S., Cronin, T. M., Dowett, H. J., Bybell, L. M., Jung, P., and Obando, J. A., 1992, Closure of the Isthmus of Panama: The near-shore marine record of Costa Rica and western Panama: Geological Society of America Bulletin, v. 104, p. 814–828.

Coates, D. A., 1969, Stratigraphy and sedimentation of the Sauce Grande Formation, Sierra de la Ventana, southern Buenos Aires Province, Argentina: Gondwana stratigraphy: Paris, United Nations Educational, Scientific, and Cultural Organization, v. 2, p. 799–819.

Coates, D. A., 1985, Late Paleozoic glacial patterns in the central Transantarctic Mountains, Antarctica, in Turner, M. D., and Splettstoesser, J. F., eds., Geology of the Central Transantarctic Mountains: Washington, D.C., American Geophysical Union, Antarctic Research Series, v. 36, p. 325–338.

Coats, R. P., 1981, Late Proterozoic (Adelaidean) tillites of the Adelaide Geosyncline, in Hambrey, M. J., and Harland, W. B., eds., Earth's pre-Pleistocene glacial record: Cambridge, United Kingdom, Cambridge University Press, p. 537–548.

Cobbing, E. J., 1981, Tillites of the base of the possibly Early Palaeozoic Marcona Formation, southwest coastal Peru, in Hambrey, M. J., and Harland, W. B., eds., Earth's pre-Pleistocene glacial record: Cambridge, United Kingdom, Cambridge University Press, p. 899–901.

Cohen, J., and Stewart, I., 1994, The collapse of Chaos: New York, Penguin Books USA, Inc., 495 p.

Coleman, A. P., 1907, A Lower Huronian Ice Age: American Journal of Science, v. 23, p. 187–192.

Coleman, A. P., 1926, Ice ages, Recent and ancient: New York, Macmillan, 296 p.

Coleman, M., and Hodges, K., 1995, Evidence for Tibetan plateau uplift before 14 Myr ago from a new minimum age for east-west extension: Nature, v. 374, p. 49–52.

Collins, L. S., Coates, A. G., Berggren, W. A., Aubry, M.-P., and Zhang, J., 1996, The late Miocene Panama isthmian strait: Geology, v. 24, p. 687–690.

Collerson, K. D., and Kamber, B. S., 1999, Evolution of the continents and the atmosphere inferred from Th-U-Nb systematics of the depleted mantle: Science, v. 283, p. 1519–1522.

Collinson, J. D., Bevins, R. E., and Clemmensen, L. B., 1989, Post-glacial mass flow and associated deposits preserved in palaeovalleys: the Late Precambrian Morænesø Formation, North Greenland: Meddelelser om Grønland: Geoscience, v. 21, 26 p.

Collinson, J. W., Isbell, J. L., Elliot, D. H., Miller, M. F., Miller, J. M. G., and Veevers, J. J., 1994, Permian-Triassic transantarctic basin, in Veevers, J. J., and Powell, C. McA., eds., Permian-Triassic Pangean basins and foldbelts along the Panthalassan margin of Gondwanaland: Geological Society of America Memoir 184, p. 173–222.

Cook, P. J., and Shergold, J. H., 1984, Phosphorous, phosphorites, and skeletal evolution at the Precambrian/Cambrian boundary: Nature, v. 308, p. 231–236.

Cooper, J. D., Droser, M. L., and Finney, S. C., eds., 1995, Ordovician Odyssey: Short papers for the Seventh International Symposium on the Ordovician System, Las Vegas, Nevada: Pacific Section Society for Sedimentary Geology (SEPM), Fullerton, California, 498 p.

Copper, P., 1977, Paleolatitudes in the Devonian of Brazil and the Frasnian-Famennian mass extinction: Palaeogeography, Palaeoclimatology, Palaeoecology, v. 21, p. 165–207.

Copper, P., 1986, Frasnian/Famennian mass extinctions and cold-water oceans: Geology, v. 14, p. 835–839.

Corbitt, L. L., and Woodward, L. A., 1973, Upper Precambrian(?) diamictite of Florida Mountains, southwestern New Mexico: Geological Society of America Bulletin, v. 84, p. 171–174.

Corfu, F., and Andrews, A. J., 1986, A U-Pb age for mineralized Nipissing diabase, Gowganda, Ontario: Canadian Journal of Earth Sciences, v. 23, p. 107–109.

Corrigan, D., and Hanmer, S., 1997, Anorthosite and related granitoids in the Grenville orogen: A product of the convective thinning of the lithosphere?: Geology, v. 25, p. 61–64.

Corsetti, F. A., and Kaufman, A. J., 1994, Chemostratigraphy of Neoproterozoic-Cambrian units, White-Inyon region, eastern California and western Nevada: Implications for global correlation and faunal distribution: Palaios, v. 9, p. 211–219.

Cortelezzi, C. R., and Solís, J., 1982, The supposed glacial sediments of the Lower Gondwana in the Aguas Blancas area (Oran, Province of Salta), Argentina, *in* Evinson, E. G., Schlüchter, Ch., and Rabassa, J., eds., Proceedings of the International Union for Quaternary Research (INQUA) Symposia on the Genesis and Lithology of Quaternary Deposits, USA, 1981/Argentina 1982: Rottterdam, Netherlands, Balkema, p. 297–300.

Cowie, J. W., and Bassett, M. G., compilers, 1989, International Union of Geological Sciences, 1989 Global Stratigraphic Chart: Supplement to Episodes, v. 12, chart.

Crawford, A. R., and Daily, B., 1971, Probable non-synchroneity of Late Precambrian glaciations: Nature, v. 230, p. 111–112.

Crittenden, M. D., Jr., Christie-Blick, N., and Link, P. K., 1983, Evidence for two pulses of glaciation during the late Proterozoic in northern Utah and southeastern Idaho: Geological Society of America Bulletin, v. 94, p. 437–450.

Crough, S. T., and Thompson, G. A., 1977, Upper mantle origin of Sierra Nevada uplift: Geology, v. 5., p. 396–399.

Crowell, J. C., 1957, Origin of pebbly mudstones: Geological Society of America Bulletin, v. 6, p. 993–1010.

Crowell, J. C., 1964, Climatic significance of sedimentary deposits containing dispersed megaclasts, *in* Nairn, A. E. R., ed., Problems in palaeoclimatology: London, John Wiley and Sons, p. 86–98, 110–111.

Crowell, J. C., 1978, Continental glaciation, cyclothems, continental positioning, and climate change: American Journal of Science, v. 278, p. 1345–1372.

Crowell, J. C., 1981, Early Paleozoic glaciation and continental drift, *in* McElhinny, M. W., and Valencio, D. A., eds., Paleoreconstructions of the continents, Geodynamic series: Washington, D.C. American Geophysical Union, v. 2, p. 45–49.

Crowell, J. C., 1983a, The recognition of ancient glaciations, *in* Medaris, L. G., Jr., Byers, C. W., Mickelson, D. M., and Shanks, W. C., eds., Proterozoic geology: Selected Papers from an International Proterozoic Symposium: Geological Society of America Memoir 161, p. 289–297.

Crowell, J. C., 1983b, Ice ages recorded on Gondwanan continents, Du Toit Memorial Lecture No. 18: Geological Society of South Africa Transactions, v. 86, p. 237–262.

Crowell, J. C., 1995, The ending of the Late Paleozoic ice age during the Permian Period, *in* Scholle, P. A., Peryt, T. M., and Ulmer-Scholle, D. S., eds., The Permian of Northern Pangea, v. 1, Paleogeography, paleoclimates, stratigraphy: Berlin, Springer-Verlag, p. 62–74.

Crowell, J. C., and Frakes, L. A., 1970, Phanerozoic glaciation and the causes of ice ages: American Journal of Science, v. 268, p. 193–224.

Crowell, J. C., and Frakes, L. A., 1971a, Late Palaeozoic glaciation of Australia: Geological Society of Australia Journal, v. 17, pt. 2, p. 115–155.

Crowell, J. C., and Frakes, L. A., 1971b, Late Paleozoic glaciation: Part IV, Australia: Geological Society of America Bulletin, v. 82, p. 2515–2540.

Crowell, J. C., and Frakes, L. A., 1972, Late Paleozoic glaciation: Part V, Karroo basin, South Africa: Geological Society of America Bulletin, v. 83, p. 2887–2912.

Crowell, J. C., and Frakes, L. A., 1975, The Late Palaeozoic glaciation, *in* Campbell, K. S. W., ed., Gondwana geology: Canberra, Australian National University Press, p. 313–331.

Crowell, J. C., Suárez-Soruco, R., and Rocha-Campos, A. C., 1980, Silurian glaciation in central South America, *in* Cresswell, M. M., and Vella, P., eds., Gondwana Five: Selected papers and abstracts of papers presented at the Fifth International Gondwana Symposium, Wellington, New Zealand: Rotterdam, Netherlands, Balkema, p. 105–110.

Crowley, T. J., and Baum, S. K., 1991a, Estimating Carboniferous sea-level fluctuations from Gondwanan ice extent: Geology, v. 19, p. 975–977.

Crowley, T. J., and Baum, S. K., 1991b, Toward reconciliation of Late Ordovician (~440 Ma) glaciation with very high CO_2 levels: Journal of Geophysical Research, v. 96, p. 22597–22610.

Crowley, T. J., and Baum, S. K., 1992, Modeling late Paleozoic glaciation: Geology, v. 20, p. 507–510.

Crowley, T. J., and Baum, S. K., 1993, Effect of decreased solar luminosity on Late Precambrian ice extent: Journal of Geophysical Research, v. 98, p. 16723–16732.

Crowley, T. J., and Baum, S. K., 1995, Reconciling Late Ordovician (440 Ma) glaciation with very high (14X) CO_2 levels: Journal of Geophysical Research, v. 100, p. 1093–1101.

Crowley, T. J., and North, G. R., 1991, Paleoclimatology: New York, Oxford University Press, and Oxford, United Kingdom, Clarendon Press, 339 p.

Crowley, T. J., Short, D. A., Mengel, J. G., and North, G. R., 1986, Role of seasonality in the evolution of climate over the last 100 million years: Science, v. 231, p. 579–584.

Crowley, T. J., Mengel, J. G., and Short, D. A., 1987, Gondwanaland's seasonal cycle: Nature, v. 329, p. 803–806.

Crowley, T. J., Baum, S. K., and Hyde, W. T., 1991, Climate model comparison of Gondwanan and Laurentide glaciations: Journal of Geophysical Research, v. 96, p. 9217–9226.

Crowley, T. J., Baum, S. K., and Hyde, W. T. 1992, Milankovitch fluctuations on supercontinents: Geophysical Research Letters, v. 19, p. 793–796.

Culver, S. J., Pojeta, J., and Repetski, J. E., 1988, First record of Early Cambrian shelly microfossils from West Africa: Geology, v. 16, p. 596–599.

Curry, W. B., and Crowley, T. J., 1987, The $\delta^{13}C$ of equatorial Atlantic surface waters: Implications for ice-age pCO_2 levels: Paleoceanography, v. 2, p. 489–517.

Daily, B., Gostin, V. A., and Nelson, C. A., 1973, Tectonic origin for an assumed glacial pavement of late Proterozoic age, South Australia: Geological Society of Australia Journal, v. 20, p. 75–78.

Dalla Salda, L., Cingolani, C., and Varela, R., 1992, Early Paleozoic orogenic belt of the Andes in southwestern South America: Result of Laurentia-Gondwana collision?: Geology, v. 20, p. 617–620.

Daly, M. C., Lawrence, S. R., Kimunía, D., and Binga, M., 1991, Late Palaeozoic deformation in central Africa: A result of distant collision?: Nature, v. 350, p. 605–607.

Dalziel, I. W. D., 1991, Pacific margins of Laurentia and East Antarctica–Australia as a conjugate rift pair: Evidence and implications for an Eocambrian supercontinent: Geology, v. 19, p. 598–601.

Dalziel, I. W. D., 1994, Precambrian Scotland as a Laurentia-Gondwana link: Origin and significance of cratonic promontories: Geology, v. 22, p. 589–592.

Dalziel, I. W. D., 1997, Neoproterozoic-Paleozoic geography and tectonics: Review, hypothesis, environmental speculation: Geological Society of America Bulletin, v. 109, p. 16–42.

Dalziel, I. W. D., and McMenamin, M. A. S., 1995, Are Neoproterozoic glacial deposits preserved on the margins of Laurentia related to the fragmentation of two supercontinents?: Comment and reply: Geology, v. 23, p. 959–961.

Dalziel, I. W. D., Dalla Salla, L. H., and Gahagan, L. M., 1994, Paleozoic Laurentia-Gondwana interaction and the origin of the Appalachian-Andean mountain system: Geological Society of America Bulletin, v. 106, p. 243–252.

Dansgaard, W., 1964, Stable isotopes in precipitation: Tellus, v. 16, p. 436–468.

Dansgaard, W., and Tauber, H., 1969, Glacial oxygen-18 content and Pleistocene ocean temperatures: Science, v. 166, p. 499–502.

Dauphin, J. P., and Simoneit, B. R. T., eds., 1991, The Gulf and Peninsular Province of the Californias: Tulsa, Oklahoma, American Association of Petroleum Geologists, 834 p.

Davidson-Arnott, R. J. D., Nickling, W., and Fahey, B. D., eds., 1982, Research in glacial, glacio-fluvial, and glacio-lacustrine systems: Proceedings of the 6th Guelph symposium on geomorphology: Norwich, United Kingdom, Geo Books, 318 p.

Davison, S., and Hambrey, M. J., 1996, Indications of glaciation at the base of the Proterozoic Stoer Group (Torridonian), NW Scotland: Geological Society of London Journal, v. 153, p. 139–149.

Davison, S., and Hambrey M. J., 1997, Indications of glaciation at the base of the Proterozoic Stoer Group (Torridonian), NW Scotland: Discussion: Geological Society of London Journal, v. 154, p. 1087–1088.

De Geer, G., 1912, A geochronology of the last 12,000 years: Comptes Rendus, Stockholm Sweden, 9th International Geological Congress, v. 1, p. 241–258.

Denton, G. H., and Hughes, T. J., eds., 1981a, The Last Great Ice Sheets: New York, John Wiley and Sons, 484 p.

Denton, G. H., and Hughes, T. J., 1981b, The Arctic ice sheets: An outrageous hypothesis, *in* Denton, G. H., and Hughes, T. J., eds., The Last Great Ice Sheets: New York, John Wiley and Sons, p. 440–467.

Derby, O. A., 1888, Uber Spuren einer carbonen Eiszeit in Sudamerika: Neues Jahrbuch für Mineralogie, Geologie und Päleontologie, v. 2, p. 171–176.

Derry, L. A., Keto, L. S., Jacobsen, S. B., Knoll, A. H., and Swett, K., 1989, Sr isotopic variations in Upper Proterozoic carbonates from Svalbard and East Greenland: Geochimica et Cosmochimica Acta, v. 53, p. 2231–2339.

Derry, L. A., Kaufman, A. J., and Jacobsen, S. B., 1992, Sedimentary cycling and environmental change in the Late Proterozoic: Evidence from stable and radiogenic isotopes: Geochimica et Cosmochimica Acta, v. 56, p. 1317–1329.

Des Marais, D. J., Strauss, H., Summons, R. E., and Hayes, J. M., 1992, Carbon isotope evidence for the stepwise oxidation of the Proterozoic environment: Nature, v. 359, p. 605–609.

de Silva, S. L., and Zielinski, G. A., 1998, Global influence of the AD 1600 eruption of Huaynaputina, Peru: Nature, v. 393, p. 455–458.

De Villiers, P. R., and Visser, J. N. J., 1977, The glacial beds of the Griqualand West Supergroup as revealed by four deep boreholes between Postmasburg and Sishen: Geological Society of South Africa Transactions, v. 80, p. 1–8.

de Wit, M. J., Roering, C., Hart, R. J., Armstrong, R. A., de Ronde, C. E. J., Green, R. W. E., Tredoux, M., Peberdy, E., and Hart, R. A., 1992, Formation of an Archaean continent: Nature, v. 357, p. 553–562.

Deynoux, M., 1980, Les formations glaciaires du Précambrian terminal et de la fin de l'Ordovicien en Afrique de l'Ouest: Deux exemples de glaciation d'inlandsis sur une plate-forme stale: Travaux des Laboratoires des Sciences de la Terre, Saint-Jérôme, Marseille b, v. 17, 554 p.

Deynoux, M., 1982, Periglacial polygonal structures and sand wedges in the Late Precambrian glacial formations of the Taoudenia Basin in Adrar of Mauritania (West Africa): Palaeogeography, Palaeoclimatology, Palaeoecology, v. 39, p. 55–70.

Deynoux, M., 1985, Terrestrial or waterlain glacial diamictites? Three case studies from the late Precambrian and Late Ordovician glacial drift in West Africa: Palaeogeography, Palaeoclimatology, Palaeoecology, v. 51, p. 97–141.

Deynoux, M., and Trompette, R., 1976, Late Precambrian mixtites: glacial and/or nonglacial? Dealing especially with the mixtites of West Africa: American Journal of Science, v. 276, p. 1302–1315.

Deynoux, M., and Trompette, R., 1981, Late Precambrian tillites of the Taoudeni Basin, West Africa, *in* Hambrey, M. J., and Harland, W. B., eds., Earth's pre-Pleistocene glacial record: Cambridge, United Kingdom, Cambridge University Press, p. 123–131.

Deynoux, M., Trompette, R., Clauer, N., and Sougy, J., 1978, Upper Precambrian and lowermost Palaeozoic correlations in West Africa and in the western part of Central Africa. Probable diachronism of the Late Precambrian

tillite: Geologische Rundschau, v. 67, p. 615–630.

Deynoux, M., Kocurek, G., and Proust, J. N., 1989, Late Proterozoic periglacial aeolian deposits on the West African Platform, Taoudeni Basin, western Mali: Sedimentology, v. 36, p. 531–549.

Deynoux, M., Proust, J. N., and Simon, B., 1991, Late Proterozoic glacially controlled shelf sequences in Western Mali (West Africa): Journal of African Earth Sciences, v. 12, p. 181–198.

Deynoux, M., Miller, J. M. G., Domack, E. W., Eyles, N., Fairchild, I. J., and Young, G. M., eds., 1994, Earth's glacial record: Cambridge, United Kingdom, Cambridge University Press, 266 p.

Díaz-Martínez, E., and Isaacson, P. E., 1995, Late Devonian glacially influenced marine sedimentation in western Gondwana: The Cumuná Formation, Altiplano, Bolivia, *in* Beauchamp, B., Embry, A. F., and Class, D., eds., Pangea: Global Environments and Resources: Carboniferous to Jurassic Pangea: Calgary, Alberta, Canadian Society of Petroleum Geologists Memoir 17, p. 511–522.

Dickins, J. M., 1976, Correlation chart for the Permian system of Australia: Canberra Australian Bureau of Mineral Resources Bulletin 156B.

Dickins, J. M., 1985, Late Palaeozoic glaciation: Australian Bureau of Mineral Resources: Research Journal of Geology and Geophysics, v. 9, p. 163–169.

Dickins, J. M., 1993, Climate of the Late Devonian to Triassic: Palaeogeography, Palaeoclimatology, Palaeoecology, v. 100, p. 89–94.

Dickens, J. M., 1997, Some problems of the Permian (Asselian) glaciation and the subsequent climate in the Permian, *in* Martini, I. P., ed., Late Glacial and Postglacial Environmental Changes: New York, Oxford University Press, p. 243–245.

Dickins, J. M., and Shah, S. C., 1987, The relationship of the Indian and Western Australian Permian marine faunas, *in* McKenzie, G. D., ed., Gondwana six: Stratigraphy, sedimentology, and paleontology: Washington, D.C., American Geophysical Union, Geophysical Monograph 41, p. 15–21.

Dickinson, W. R., Soreghan, G. S., and Giles, K. A., 1994, Glacio-eustatic origin of Permo-Carboniferous stratigraphic cycles: Evidence from the southern Cordilleran foreland region, *in* Tectonics and eustatic controls on sedimentary cycles: SEPM Concepts in sedimentology, v. 4, p. 25–34.

Doré, F., 1981, Late Precambrian tilloids of Normandy (Armorican Massif), *in* Hambrey, M. J., and Harland, W. B., eds., Earth's pre-Pleistocene glacial record: Cambridge, United Kingdom, Cambridge University Press, p. 643–646.

Doré, F., Dupret, L., and Le Gall, J., 1985, Tillites et tilloides du Massif Armoricain: Palaeogeography, Palaeoclimatology, Palaeoecology, v. 51, p. 85–96.

Dos Santos, P. R., Rocha-Campos, A. C., and Canuto, J. R., 1996, Patterns of Palaeozoic deglaciation in the Paraná Basin, Brazil: Palaeogeography, Palaeoclimatology, Palaeoecology, v. 125, p. 165–184.

Dott, R. H., Jr., 1961, Squantum 'tillite,' Massachusetts—evidence of glaciation or subaqueous mass movement?: Geological Society of America Bulletin, v. 71, p. 1289–1306.

Dott, R. H., Jr., 1963, Dynamics of subaqueous gravity depositional processes: American Association of Petroleum Geologists Bulletin, v. 47, p. 104–128.

Douglas, R. G., and Woodruff, F., 1981, Deep sea benthonic foraminifera, *in* Emiliani, C., ed., The Sea, v. 7: New York, Wiley-Interscience, p. 1233–1327.

Dow, D. B., Beyth, M., and Hailu, T., 1971, Palaeozoic glacial rocks recently discovered in northern Ethiopia: Geological Magazine, v. 108, p. 53–59.

Dowdeswell, J. A., and Scourse, J. D., eds., 1990, Glacimarine environments: Processes and sediments: Geological Society of London Special Publication 53, 423 p.

Drake, D., 1815, Natural and statistical view or picture of Cincinnati and the Miami country: Cincinnati, Ohio, Looker and Wallace, 251 p.

Drake, D., 1825, Geological account of the valley of the Ohio: American Philosophical Society Transactions, v. 2, p. 124–139.

Drake, L. D., 1971, Evidence for ablation and basal till in east-central New Hampshire, *in* Goldthwait, R. P., ed., Till: a symposium: Columbus, Ohio State University, p. 13–91.

Drake, L. D. 1972, Mechanisms of clast attrition in basal till: Geological Society of America Bulletin, v. 83, p. 2159–2165.

Dreimanis, A., 1983, Quaternary glacial deposits: Implications for the interpretation of Proterozoic glacial deposits, *in* Medaris, L. G., Jr., Byers, C. W., Mickelson, D. M., and Shanks, W. C., eds., Proterozoic geology: Selected Papers from an International Proterozoic Symposium: Geological Society of America Memoir 161, p. 299–307.

Drewry, D., 1986, Glacial geologic processes: London, Edward Arnold, 276 p.

Drexel, J. F., Preiss, W. V., and Parker, A. J., eds., 1993, Geology of South Australia, v. 1, The Precambrian: Geological Survey of South Australia, Bulletin 54, p. 170–203.

Drinkwater, N. J., Pickering, K. T., and Siedlecka, A., 1996, Deep-water fault-controlled sedimentation, Arctic Norway and Russia: Response to Late Proterozoic rifting and the opening of the Iapetus Ocean: Geological Society of London Journal, v. 153, p. 427–436.

Driscoll, N. W., and Haug, G. H., 1998, A short circuit in thermohaline circulation: A cause for Northern Hemisphere glaciation? Science, v. 282, p. 436–438.

Duff, P. McL. D., Hallam, A., and Walton, E. K., 1967, Cyclic sedimentation: Amsterdam, Developments in sedimentology, v. 10: Amsterdam, Elsevier, 280 p.

Dürr, S. B., and Dingeldey, 1996, The Koko belt (Namibia): Part of a late Neoproterozoic continental-scale strike-slip system: Geology, v. 24, p. 503–506.

Du Toit, A. L., 1921, The Caboniferous glaciation of South Africa: Geological Society of South Africa Transactions, v. 24, p. 188–277.

Du Toit, A. L., 1937, Our wandering continents: Edinburgh, Oliver and Boyd, 366 p. (reprinted, 1957, New York, Hafner).

Du Toit, A. L., 1954, Geology of South Africa: Edinburgh, Oliver and Boyd, (3rd ed.), 611 p.

Eddy, J. A., Gilliland, R. L., and Hoyt, D. V., 1982, Changes in the solar constant and climate effects: Nature, v. 300, p. 689–694.

Edmond, J. M., 1992, Himalayan tectonics, weathering processes, and the strontium isotope record in marine limestones: Science, v. 258, p. 1594–1597.

Edwards, M. B., 1984, Sedimentology of the Upper Proterozoic glacial record, Vestertana Group, Finnmark, North Norway: Norges Geologiske Undersøkelse, Bulletin 394, 76 p.

Edwards, M. B., 1997, Discussion of glacial or non-glacial origin for the Bigganjargga tillite, Finnmark, northern Norway: Discussion: Geological Magazine, v. 134, p. 873–876.

Edwards, M. B. and Føyn, S., 1981, Late Precambrian tillites in Finnmark, North Norway, *in* Hambrey, M. J., and Harland, W. B., Earth's pre-Pleistocene glacial record: Cambridge, United Kingdom, Cambridge University Press, p. 606–610.

Eickhoff, K.-H., Von der Borch, C. C., and Grady, A. E., 1988, Proterozoic canyons of the Flinders Ranges (South Australia): submarine canyons or drowned river valleys?: Sedimentary geology, v. 58, p. 217–235.

Einsele, G., and Seilacher, A., eds., 1982, Cyclic and event stratification: Berlin, Springer-Verlag, 536 p.

Eisbacher, G. H., 1981, Sedimentary tectonics and glacial record in the Windermere Supergroup, Mackenzie Mountains, Northwestern Canada: Ottawa Geological Survey of Canada Paper 80-27, 40 p.

Eisbacher, G. H., 1985, Late Proterozoic rifting, glacial sedimentation, and sedimentary cycles in the light of Windermere deposition, Western Canada: Palaeogeography, Palaeoclimatology, Palaeoecology, v. 51, p. 231–254.

Elliot, D. N., 1975, Gondwana basins of Antarctica, *in* Campbell, K. S. W., ed., Gondwana Geology: Canberra, Australia, Australian National University Press, p. 493–536.

El-Nakhal, H. A., 1984, Possible Late Palaeozoic glaciation in the central parts of the Yemen Arab Republic: Journal of Glaciology, v. 30, p. 126–128.

Elston, D. P., Link, P. K., Winston, D., and Horodyski, R. J., 1993, Correlations of middle and late Proterozoic successions, *in* Link, P. K., Christie-Blick, N., Devlin, W. J., Elston, D. P., Horodyski, R. J., Levy, M., Miller, J. M. G., Pearson, R. C., Prave, A., Stewart, J. H., Winston, D., Wright, L. A., and Wrucke, C. T., 1993, Middle and Late Proterozoic stratified rocks of the western U.S. Cordillera, Colorado Plateau, and Basin and Range province, *in* Reed, J. C., Jr., Bickford, M. E., Houston, R. S., Link, P. K., Rankin, D. W., Sims, P. K., and Van Schmus, W. R., eds., Precambrian: Conterminous U.S.: Boulder, Colorado, Geological Society of America, The Geology of North America, v. C-2, p. 463–595.

Embleton, B. J. J., 1984, Australia's global setting: Past global settings, *in* Veevers, J. J., ed., Phanerozoic earth history of Australia, Oxford geology series, v. 2: Oxford, United Kingdom, Clarendon Press, p. 11–17.

Embleton, B. J. J., and Williams, G. E., 1986, Low palaeolatitude of deposition for late Precambrian periglacial varvites in South Australia: implications for palaeoclimatology: Earth and Planetary Science Letters, v. 79, p. 419–430.

Embleton, C., and King, C. A. M., 1975a, Periglacial Geomorphology: New York, Halsted Press, John Wiley and Sons, 203 p.

Embleton, C., and King, C. A. M., 1975b, Glacial geomorphology: New York, Halsted Press, John Wiley and Sons, 573 p.

Emery, K. O., 1955, Transportation of rocks by driftwood: Journal of Sedimentary Petrology, v. 25, p. 51–57.

Emery, K. O., 1963, Organic transportation of marine sediments, *in* Hill, M. N., ed., The Sea, v. 3: New York, Wiley, p. 776–793.

Emiliani, C., 1955, Pleistocene temperatures: Journal of Geology, v. 63, p. 538–578.

Epshteyn, O. G., 1978, Mesozoic-Cenozoic climates of northern Asia and glacial marine deposits: International Geology Review, v. 20, p. 49–58.

Erdtmann, B.-D., Kley, J., Müller, J., and Jacobshagen, V., 1995, Ordovician basin dynamics and new graptolite data from the Tarija Region, Eastern Cordillera, South Bolivia, *in* Cooper, J. D., Roser, M. L., and Finney, S. C., eds., Ordovician Odyssey: Short papers for the Seventh International Symposium on the Ordovician System, Las Vegas, Nevada: Fullerton, California, Society of Economic Paleontologists and Mineralogists, p. 69–73.

Ernstson, K., and Claudin, F., 1990, Pelarda Formation (Eastern Iberian Chains, NE Spain): Ejecta of Azuara impact structure: Neus Jahrbuch für Geologie und Paläontologie Monashefte, v. 10, p. 581–599.

Erwin, D. H., 1993, The great Paleozoic crisis: Life and death in the Permian: New York, Columbia University Press, 327 p.

Erwin, D. H., 1994, The Permo-Triassic extinction: Nature, v. 367, p. 231–236.

Evans, D. A., 1998, True polar wander, a supercontinent legacy: Earth and Planetary Science Letters, v. 157, p. 1–8.

Evans, D. A., Beukes, N. J., and Kirschvink, J. L., 1997, Low-latitude glaciation in the Palaeoproterozoic era: Nature, v. 386, p. 262–266.

Evans, J. A., Fitches, W. R., and Muir, R. J., 1998, Laurentian clasts in a Neoproterozoic tillite from Scotland: Journal of Geology, v. 106, p. 361–366.

Evans, J. V., 1982, The Sun's influence on the Earth's atmosphere and interplanetary space: Science, v. 216, p. 467–474.

Evans, R. H. S., and Tanner, P. W. G., 1996, A late Vendian age for the Kinlochlaggan boulder bed (Dalradian)?: Geological Society of London Journal, v. 153, p. 823–826.

Eyles, C. H., 1988, Glacially and tidally influenced shallow marine sedimentation of the Late Precambrian Port Askaig Formation, Scotland: Palaeogeography, Palaeoclimatology, Palaeoecology, v. 68, pp. 1–25.

Eyles, C. H., and Eyles, N., 1983, Glaciomarine model for upper Precambrian diamictites of the Port Askaig Formation, Scotland: Geology, v. 11, p. 692–696.

Eyles, C. H., and Lagoe, M. B., 1990, Sedimentation patterns and facies geometries on a temperate glacially-influenced continental shelf: the Yakataga Formation, Middleton Island, Alaska. *in* Dowdeswell, J. A., and Scourse, J. D., eds., Glacimarine environments: processes and sediments: Geological Society of London Special Publication 53, p. 363–386.

Eyles, C. H., Eyles, N., and Lagoe, M. B., 1991, The Yakataga Formation: a late Miocene to Pleistocene record of temperate glacial marine sedimentation in the Gulf of Alaska, *in* Anderson, J. B., and Ashley, G. M., eds., Glacial marine sedimentation: Paleoclimatic significance: Geological Society of America Special Paper 261, p. 159–180.

Eyles, C. H., Eyles, N., and França, A. B., 1993, Glaciation and tectonics in an active intracratonic basin: the Late Palaeozoic Itararé Group, Paraná Basin, Brazil: Sedimentology, v. 40, p. 1–25.

Eyles, N., 1990, Marine debris flows: Late Precambrian útillitesî of the Avalonian-Cadomian belt: Palaeogeography, Palaeoclimatology, Palaeoecology, v. 79, p. 73–98.

Eyles, N., 1993, Earth's glacial record and its tectonic setting: Earth-Science Reviews, v. 35, 248 p.

Eyles, N., 1996, Passive margin uplift around the North Atlantic region and its role in Northern Hemisphere late Cenozoic glaciation: Geology, v. 24, p. 103–106.

Eyles, N., and Clark, B. M., 1985, Gravity-induced soft-sediment deformation in glaciomarine sequences of the Upper Proterozoic Port Askaig Formation, Scotland: Sedimentology, v. 32, p. 784–814.

Eyles, N., and Eyles, C. H., 1989, Glacially-influenced deep-marine sedimentation of the Late Precambrian Gaskiers Formation, Newfoundland, Canada: Sedimentology, v. 36, p. 601–620.

Eyles, N., and Eyles, C. H., 1993, Glacial geologic confirmation of an intraplate boundary in the Paraná basin of Brazil: Geology, v. 21, p. 459–462.

Eyles, N., and Young, G. M., 1994, Geodynamic controls on glaciation in Earth history, *in* Deynoux, M., Miller, J. M. G., Domack, E. W., Eyles, N., Fairchild, I. J., and Young, G. M., eds.: Cambridge, United Kingdom, Cambridge University Press, p. 1–28.

Fairbanks, R. G., and Matthews, R. K., 1978, The marine oxygen isotope record in Pleistocene coral, Barbados, West Indies: Quaternary Research, v. 10, p. 181–196.

Fairchild, I. J., 1993, Balmy shores and icy wastes: the paradox of carbonates associated with glacial deposits in Neoproterozoic times: Sedimentology Review, v. 1, p. 1–16.

Fairchild, I. J., and Hambrey, M. J., 1995, Vendian basin evolution in East Greenland and NE Svalbard: Precambrian Research, v. 73, p. 217–233.

Fairchild, I. J., and Spiro, B., 1990, Carbonate minerals in glacial sediments: geochemical clues to palaeoenvironment, *in* Dowdeswell, J. A., and Scourse, J. D., eds., Glacimarine environments: processes and sediments: Geological Society of London Special Publication 53, p. 201–216.

Fairchild, I. J., Hambrey, M. J., Spiro, J., and Jefferson, T. H., 1989, Late Proterozoic glacial carbonates in northeast Spitzbergen: New insight into the carbonate-tillite association: Geological Magazine, v. 126, p. 469–490.

Fairchild, I. J., Bradby, L., and Spiro, B., 1994, Reactive carbonate in glacial systems: a preliminary synthesis of its creation, dissolution and reincarnation, *in* Deynoux, M., Miller, J. M. G., Domack, E. W., Eyles, N., Fairchild, I. J., and Young, G. M., eds.: Cambridge, United Kingdom, Cambridge University Press, p. 176–192.

Falkowski, P. G., Kim, Y., Kolber, Z., Wilson, C., Wirick, C., and Cess, R., 1992, Natural versus anthropogenic factors affecting low-level cloud albedo over the North Atlantic: Science, v. 256, p. 1311–1313.

Fanning, C. M., Ludwig, K. R., Forbes, B. G., and Preiss, W. V., 1986, Single and multiple grain U-Pb zircon analyses for the early Adelaidean Rook Tuff, Willouran Ranges, South Australia: Geological Society of Australia Journal, Abstract, v. 15, p. 71–72.

Farley, K. A., 1995, Cenozoic variations in the flux of interplanetary dust recorded by ^{3}He in a deep-sea sediment: Nature, v. 376, p. 153–156.

Farley, K. A., and Patterson, D. G., 1995, A 100-kyr periodicity in the flux of extraterrestrial ^{3}He to the sea floor: Nature, v. 378, p. 600–603.

Faure, G., 1986, Principles of isotope geology, 2nd ed.: New York, John Wiley and Sons, 589 p.

Fedo, C. M., Nesbitt, H. W., and Young, G. M., 1995, Unraveling the effects of potassium metasomatism in sedimentary rocks and paleosols, with implications for paleoweathering conditions and provenance: Geology, v. 23, p. 921–924.

Fedo, C. M., Young, G. M., and Nesbitt, H. W., 1997, Paleoclimatic control on the composition of the Paleoproterozoic Serpent Formation, Huronian Supergroup, Canada: A greenhouse to icehouse transition: Precambrian Research, v. 86, p. 201–223.

Fernandez Garrasino, C. A., 1981, Late Palaeozoic Tarija Formation, southern Bolivia and northern Argentina, *in* Hambrey, M. J., and Harland, W. B., eds., Earth's pre-Pleistocene glacial record: Cambridge, United Kingdom, Cambridge University Press, p. 853–854.

Fiege, K., 1978, Cyclic sedimentation, *in* Fairbridge, R. W., and Bourgeois, J., Encyclopedia of sedimentology: Stroudsburg, Pennsylvania, Dowden, Hutchinson and Ross, Inc., p. 223–232.

Finney, S. C., Berry, W. B. N., Cooper, J. D., Ripperdan, R. L., Sweet, W. C., Jacobson, S. R., Soufiane, A., Achab, A., and Noble, P. J., 1999, Late

Ordovician extinction: A new perspective from stratigraphic sections in central Nevada: Geology, v. 27, p. 215–218.

Fischer, A. G., 1982, Long-term climatic oscillations recorded in stratigraphy, *in* Climate in Earth History, Studies in Geophysics: Washington, D.C., National Academy Press, p. 97–104.

Fischer, A. G., 1986, Climatic rhythms recorded in strata: Annual review of earth and planetary sciences, v. 14, p. 351–376.

Fischer, A. G., and Arthur, M. A., 1977, Secular variations in the pelagic realm, *in* Cook, H. E., and Enos, P., eds., Deep Water Carbonate Environments: Society of Economic Paleontologists and Sedimentologists Special Publication 25, p. 18–50.

Fischer, A. G., and Schwarzacher, W., 1984, Cretaceous bedding rhythms under orbital control?, *in* Berger, A., Imbrie, J., Hays, J., Kuklan, G., and Saltzman, B., eds., Milankovitch and Climate: Dordrecht, Netherlands, Reidel, p. 163–175.

Fitzgerald, P. G., and Stump, E., 1991, Early Cretaceous uplift in the Ellsworth Mountains of West Antarctica: Science, v. 354, p. 92–94.

Fleming, J. R., 1992, T. C. Chamberlin and H_2O climate feedbacks: A voice from the past: EOS, (American Geophysical Union Transactions), v. 73, p. 505–507.

Fleming, J. R., 1998, Historical perspectives on climate change: New York, Oxford University Press, 194 p.

Flint, R. F., 1965, Introduction: historical perspectives, *in* Wright, H. E., Jr., and Frey, D. G., eds., The Quaternary of the United States: Princeton, New Jersey, Princeton University Press, p. 3–11.

Flint, R. F., 1971, Glacial and Quaternary geology: New York, John Wiley and Sons, 892 p.

Flint, R. F., Sanders, J. E., and Rodgers, J., 1960, Diamictite, a substitute for symmictite: Geological Society of America Bulletin, v. 71, p. 1809.

Folk, R. L., 1975, Glacial deposits identified by chattermark trails in detrital garnets: Geology, v. 3, p. 473–475.

Föllmi, K. B., 1995, 160 m.y. record of marine sedimentary phosphorous burial: Coupling of climate and continental weathering under greenhouse and icehouse conditions: Geology, v. 23, p. 859–862.

Fortey, A. R., 1984, Global earlier Ordovician transgressions and regressions and their biological implications, *in* Bruton, D. L., ed., Aspects of Ordovician system, IV: University of Oslo Palaeontological Contributions, no. 295, p. 35–50.

Fortuin, A. R., 1984, Late Ordovician glaciomarine deposits (Orea Shale) in the Sierra De Albarracin, Spain: Palaeogeography, Palaeoclimatology, Palaeoecology, v. 48, p. 245–261.

Føyn, S., 1937, The Eo-Cambrian series of the Tana district, Northern Norway: Norsk Geologisk Tidskrift, v. 17, p. 65–164.

Føyn, S., 1985, The Late Precambrian in northern Scandinavia, *in* Gee, D. G., and Sturt, B. A., eds., The Caledonide Orogen—Scandinavia and related areas: New York, John Wiley and Sons, p. 233–245.

Frakes, L. A., 1979, Climates throughout geologic time: Amsterdam, Elsevier, 310 p.

Frakes, L. A., 1986, Mesozoic-Cenozoic climate history and causes of the glaciation, *in* Hsü, K. J., ed., Mesozoic and Cenozoic oceans, Geodynamic Series: Washington, D.C., American Geophysical Union, p. 33–48.

Frakes, L. A., and Crowell, J. C., 1967, Facies and paleogeography of Late Paleozoic Lafonian diamictite, Falkland Islands: Geological Society of America Bulletin, v. 78, p. 37–58.

Frakes, L. A., and Crowell, J. C., 1969, Late Paleozoic glaciation: I, South America: Geological Society of America Bulletin, v. 80, p. 1007–1042.

Frakes, L. A., and Crowell, J. C., 1970, Late Paleozoic glaciation: II, Africa exclusive of the Karroo Basin: Geological Society of America Bulletin, v. 81, p. 2261–2286.

Frakes, L. A., and Francis, J. E., 1988, A guide to Phanerozoic cold polar climates from high-latitude ice-rafting in the Cretaceous: Nature, v. 335, p. 547–549.

Frakes, L. A., and Francis, J. E., 1990, Cretaceous paleoclimates, *in* Ginsburg, R. N., and Beaudoin, B., eds., Cretaceous resources, events and rhythms: Dordrecht, Netherlands, Kluwer Academic Publishers, p. 273–287.

Frakes, L. A., and Kemp, E. M., 1972, Influence of continental positions on early

Tertiary climates: Nature, v. 240, p. 97–100.

Frakes, L. A., and Krassay, A. A., 1992, Discovery of probable ice-rafting in the Late Mesozoic of the Northern Territory and Queensland: Australian Journal of Earth Sciences, v. 39, p. 115–119.

Frakes, L. A., Matthews, J. L., and Crowell, J. C., 1971, The late Paleozoic glaciation: Part III, Antarctica: Geological Society of America Bulletin, v. 82, p. 1581–1604.

Frakes, L. A., Kemp, E. M., and Crowell, J. C., 1975, Late Paleozoic glaciation: Part VI, Asia: Geological Society of America Bulletin, v. 86, p. 454–464.

Frakes, L. A., Kemp, E. M., and Crowell, J. C., 1976, Late Paleozoic glaciation: Part VI, Asia: Reply: Geological Society of America Bulletin, v. 87, p. 640.

Frakes, L. A., Francis, J. E., and Syktus, J. I., 1992, Climate modes of the Phanerozoic: Cambridge, England, Cambridge University Press, 274 p.

Frakes, L. A., Alley, N. F., and Deynoux, M., 1995, Early Cretaceous ice rafting and climate zonation in Australia: International Geology Review, v. 37, p. 567–583.

França, A. B., 1994, Itararé Group: Gondwanan Carboniferous-Permian of the Paraná Basin, Brazil, *in* Deynoux, M., Miller, J. M. G., Domack, E. W., Eyles, N., Fairchild, I. J., and Young, G. M., eds., Earth's glacial record: Cambridge, United Kingdom, Cambridge University Press, p. 70–82.

França, A. B., and Potter, P. E., 1991, Stratigraphy and reservoir potential of glacial deposits of the Itararé group (Carboniferous-Permian), Paraná Basin, Brazil: American Association of Petroleum Geologists Bulletin, v. 75, p. 62–85.

France-Lanord, C., and Derry, L. A., 1997, Organic carbon burial forcing of the carbon cycle from Himalayan erosion: Nature, v. 390, p. 65–67.

Francis, J. E., 1993, Cretaceous climates: Sedimentology review, v. 1, p. 17–30.

Frank, T. D., and Lyons, T. W., 1998, "Molar-tooth" structures: A geochemical perspective on a Proterozoic enigma: Geology, v. 26, p. 683–686.

Friedman, G. M., and Sanders, J. E., 1978, Principles of sedimentology: New York, John Wiley and Sons, 792 p.

Frimmel, H. E., Klötzli, U. S., and Siegfried, P. R., 1996, New Pb-Pb single zircon age constraints on the timing of Neoproterozoic glaciation and continental break-up in Namibia: Journal of Geology, v. 104, p. 459–469.

Gagnier, P. Y., Blieck, A., Emig, C. C., Sempere, T., Vachard, D., and Vanguestaine, M., 1996, New paleontological data on the Ordovician and Silurian of Bolivia: Journal of South American Earth Sciences, v. 9, p. 329–347.

Gair, J. E., 1981, Lower Proterozoic glacial deposits of northern Michigan, U.S.A., *in* Hambrey, M. J., and Harland, W. B., eds., Earth's pre-Pleistocene glacial record: Cambridge, United Kingdom, Cambridge University Press, p. 803–806.

Ganopolski, A., Rahmstorf, S., Petoukhov, V., and Claussen, M., 1998, Simulation of modern and glacial climates with a coupled global model of intermediate complexity: Nature, v. 391, p. 351–356.

Garwood, E. J., and Gregory, J. W., 1898, Contribution to the glacial geology of Spitzbergen: Quarterly Journal of Geological Society of London, v. 54, p. 197–225.

Gayer, R. A., and Rice, A. H. N., 1989, Palaeogeographic reconstruction of the pre- to syn-Iapetus rifting sediments in the Caledonides of Finnmark, N. Norway, *in* Gayer, R. A., ed., The Caledonide geology of Scandinavia: London, Graham and Trotman, p. 127–142.

Geldsetzer, H. H. J., Goodfellow, W. D., McLaren, D. J., and Orchard, M. J., 1987, Sulfur-isotope anomaly associated with the Frasnian-Famennian extinction, Medicine Lake, Alberta, Canada: Geology, v. 15, p. 393–396.

Geldsetzer, H. H. J., Goodfellow, W. D., McLaren, D. J., and Orchard, M. J., 1988, Sulfur-isotope anomaly associated with the Frasnian-Famennian extinction, Medicine Lake, Alberta, Canada: Reply: Geology, v. 16, p. 87–88.

Gensel, P. G., and Andrews, H. N., 1987, The evolution of early land plants: American Scientist, v. 75, p. 478–489.

Germs, G. J. B., 1995, The Neoproterozoic of southwestern Africa, with emphasis on platform stratigraphy and paleontology: Precambrian Research, v. 73, p. 137–151.

Gibbs, M. T., Barron, E. J., and Kump, L. R., 1997, An atmospheric pCO_2 threshold for glaciation in the Late Ordovician: Geology, v. 25, p. 447–450.

Gilbert, R., 1990, Rafting in glacimarine environments, *in* Dowdeswell, J., and Scourse, J. D., eds., Glacimarine environments: processes and sediments:

Geological Society of London Special Publication 53, p. 105–120.

Goddard, A. B., 1992, The deposition style and tectonic setting of the Early Proterozoic Turee Creek Group in the Hardey Syncline, Hamersley Province, Northwestern Australia [Honours thesis]: Perth, University of Western Australia, 94 p.

Gold, D. J. C., and Von Veh, M. W., 1995, Tectonic evolution of the Late Archaean Pongola-Mozaan basin, South Africa: Journal of African Earth Sciences, v. 21, p. 201–212.

Goldthwait, R. P., ed., 1971, Till: a symposium: Columbus, Ohio State University Press, 402 p.

Goldthwait, R. P., ed., 1975, Glacial deposits, Benchmark paper in geology: Stroudsburg, Pennsylvania, Dowden, Hutchinson and Ross, Inc., v. 21, 464 p.

González-Bonorino, G., and Eyles, N., 1995, Inverse relation between ice extent and the late Paleozoic glacial record of Gondwana: Geology, v. 23, p. 1015–1018.

Goodwin, A. M., 1991, Precambrian geology: London, Academic Press, 666 p.

Gostin, V. A., Haines, P. W., Jenkins, R. J. F., Compston, W., and Williams, I. S., 1986, Impact ejecta horizon within late Precambrian shales, Adelaide Geosyncline, South Australia: Science, v. 233, p. 198–200.

Gower, C. F., Ryan, A. B., and Rivers, T., 1990, Mid-Proterozoic Laurentia-Baltica: An overview of its geological evolution and a summary of the contributions made in this volume, *in* Gower C. F., Rivers, T., and Ryan, A. B., eds., Mid-Proterozoic Lauentia-Baltica: Montreal, Quebec, Geological Association of Canada Special Paper 38, p. 1–20.

Gradstein, F. M., and Ogg, J., 1996, A Phanerozoic time scale: Episodes, v. 19, p. 3–4, charts.

Grahn, Y., and Caputo, M. V., 1992, Early Silurian glaciations in Brazil: Palaeogeography, Palaeoclimatology, Palaeoecology, v. 99, p. 9–15.

Grahn, Y., and Caputo, M. V., 1994, Late Ordovician evolution of the intracratonic basins in north-west Gondwana: Geologische Rundschau, v. 83, p. 665–668.

Graindor, M. J., 1965, Les tillites centré-cambriennes de Normandie: Geologische Rundschau, v. 54, p. 61–82.

Grant, W. B., Fishman, J., Browell, E. V., Brackett, V. G., Nganga, D., Minga, A., Cros, B., Veiga, R. E., Butler, C. F., Fenn, M. A. and Nowicki, G. D., 1992, Observations of reduced ozone concentrations in the tropical stratosphere after the eruption of Mt. Pinatubo: Geophysical research letters, v. 19, p. 1109–1112.

Gravenor, C. P., 1979, The nature of the Late Paleozoic glaciation in Gondwana as determined from an analysis of garnets and other heavy minerals: Canadian Journal of Earth Sciences, v. 16, p. 1137–1153.

Gravenor, C. P., 1985, Chattermarked garnets found in soil profiles and beach environments: Sedimentology, v. 32, p. 295–306.

Gravenor, C. P., and Leavitt, R. K., 1981, Experimental formation and significance of etch patterns on detrital grains: Canadian Journal of Earth Sciences, v. 18, p. 765–775.

Greenway, M. E., 1972, The geology of the Falkland Islands: British Antarctic Survey Scientific Report 76, 42 p.

Gregor, C. B., Garrels, R. M., Mackenzie, F. T., and Maynard, J. B., eds., 1988, Chemical cycles in the evolution of the Earth: New York, John Wiley and Sons, 276 p.

Gregory, R. T., Douthitt, C. B., Duddy, I. R., Rich, P. V., and Rich, T. H., 1989, Oxygen isotope composition of carbonate concretions from the Lower Cretaceous of Victoria, Australia: Implications for the evolution of meteoric waters on the Australian continent in a paleopolar environment: Earth and Planetary Science Letters, v. 92, p. 27–42.

Grey, K., and Corkeron, M., 1998, Late Neoproterozoic stromatolites in glacigenic successions of the Kimberley region, Western Australia: Evidence for a younger Marinoan glaciation: Precambrian Research, v. 92, p. 65–87.

Grieve, R., Rupert, J., Smith, J., and Therriault, A., 1995, The record of terrestrial impact cratering: GSA Today, v. 5, p. 189–196.

Grimm, E. C., Jacobson, G. L., Jr., Watts, W. A., Hansen, B. C. S., and Maasch, K. A., 1993, A 50,000-year record of climate oscillations from Florida and its temporal correlation with the Heinrich events: Science, v. 261,

p. 198–200.

Grunow, A. W., Kent, D. V., and Dalziel, I. W. D., 1991, New paleomagnetic data from Thurston Island: Implications for the tectonics of West Antarctica and Weddell Sea opening: Journal of Geophysical Research, v. 96, p. 17935–17954.

Grunow, A., Hanson, R., and Wilson, T., 1996, Were aspects of Pan-African deformation linked to Iapetus opening?: Geology, v. 24, p. 1063–1066.

Gruszczyński, Stanislaw H., Hoffman, A., and Małkowski, K., 1989, A brachiopod calcite record of the oceanic carbon and oxygen isotope shifts at the Permian/Triassic transition: Nature, v. 337, p. 64–68.

Guan Baode, Wu Ruitang, Hambrey, M. J., and Geng Wuchen, 1986, Glacial sediments and erosional pavements near the Cambrian-Precambrian boundary in western Henan Province, China: Geological Society of London Journal, v. 143, p. 311–323.

Guoqiu Gao, 1993, The temperature and oxygen-isotope composition of early Devonian oceans: Nature, v. 361, p. 712–714.

Gurnis, M., 1990, Ridge spreading, subduction, and sea level fluctuations: Science, v. 250, p. 970–972.

Gurnis, M., 1993, Phanerozoic marine inundation of continents driven by dynamic topography above subducting slabs: Nature, v. 364, p. 589–593.

Gurnis, M., and Torsvik, E. H., 1994, Rapid drift of large continents during the late Precambrian and Paleozoic: Paleomagnetic constraints and dynamic models: Geology, v. 22, p. 1023–1026.

Gutzmer, J., and Beukes, N. J., 1998, Earliest laterites and possible evidence for terrestrial vegetation in the Early Proterozoic: Geology, v. 26, p. 263–266.

Hambrey, M. J., 1981a, Palaeozoic tillites in northern Ethiopia, *in* Hambrey, M. J., and Harland, W. B., eds., Earth's pre-Pleistocene glacial record: Cambridge, United Kingdom, Cambridge University Press, p. 38–40.

Hambrey, M. J., 1981b, Late Palaeozoic Karoo tillites in Gabon, *in* Hambrey, M. J., and Harland, W. B., eds., Earth's pre-Pleistocene glacial record: Cambridge, United Kingdom, Cambridge University Press, p. 41–42.

Hambrey, M. J., 1985, The late Ordovician–early Silurian glacial period: Palaeogeography, Palaeoclimatology, Palaeoecology, v. 51, p. 273–289.

Hambrey, M. J., and Harland, W. B., eds., 1981, Earth's pre-Pleistocene glacial record: Cambridge, United Kingdom, Cambridge University Press, 1004 p.

Hambrey, M. J., and Harland, W. B., 1985, The late Proterozoic glacial era: Palaeogeography, Palaeoclimatology, Palaeoecology, v. 51, p. 255–272.

Hambrey, M. J., and Moncrieff, A. C. M., 1985, Vendian stratigraphy and sedimentology of the East Greenland Caledonides: Rapport Grønlands Geologiske Undersøgelse, v. 321, p. 88–94.

Hambrey, M. J., and Spencer, A. A., 1987, Late Precambrian glaciation of central East Greenland: Meddelelser om Grønland, Geoscience, v. 19, p. 1–50.

Hambrey, M. J., Harland, W. B., and Waddams, P., 1981, Late Precambrian tillites of Svalbard, *in* Hambrey, M. J., and Harland, W. B., eds., Earth's pre-Pleistocene glacial record: Cambridge, United Kingdom, Cambridge University Press, p. 592–600.

Hansen, J. E., and Lacis, A. A., 1990, Sun and dust versus greenhouse gases: an assessment of their relative roles in global climate change: Nature, v. 346, p. 713–719.

Hansen, V. L., Goodge, J. W., Keep, M., and Oliver, D. H., 1993, Asymmetric rift interpretation of the western North American margin: Geology, v. 21, p. 1067–1070.

Hanson, R. E., Wilson, T. J., and Wardlaw, M. S., 1988, Deformed batholiths in the Pan-African Zambezi belt, Zambia: Age and implications for regional Proterozoic tectonics: Geology, v. 16, p. 1134–1137.

Haq, B. U., Hardenbol, J., and Vail, P. R., 1988, Mesozoic and Cenozoic Chronostratigraphy and cycles of sea-level change, *in* Wilgus, C. K., Hastings, B. S., Kendall, C. G. St. C., Posamentier, H. W., Ross, C. A., and Van Wagoner, J. C., eds., Sea-level changes: Society of Economic Paleontologists and Mineralogists Special Publication 42, p. 71–108.

Harker, R. I., and Giegengack, R. F., 1989, Brecciation of clasts in diamictites of the Gowganda Formation, Ontario, Canada: Geology, v. 17, p. 123–126.

Harland, W. B., 1964, Critical evidence for a great infra-Cambrian glaciation: Geologische Rundschau, v. 54, p. 45–91.

Harland, W. B., 1972, The Ordovician ice age: Geological Magazine, v. 109,

p. 451–456.

Harland, W. B., 1981, Chronology of Earth's glacial and tectonic record: Geological Society of London Journal, v. 138, p. 197–203.

Harland, W. B., 1983, The Proterozoic glacial record, *in* Medaris, L. G., Jr., Byers, C. W., Mickelson, D. M., and Shanks, W. C., eds., Proterozoic geology: Selected papers from an international Proterozoic symposium: Geological Society of America Memoir 161, p. 279–288.

Harland, W. B., and Bidgood, D. E. T., 1959, Palaeomagnetism in some Norwegian sparagmites and the late pre-Cambrian ice age: Nature, v. 184, p. 1860–1862.

Harland, W. B., and Herod, K. N., 1975, Glaciations through time, *in* Wright, A. E., and Moseley, F., eds., Ice ages: ancient and modern: Liverpool, Seel House Press, Geological Journal Special Issue No. 6, p. 169–216.

Harland, W. B., Armstrong, R. L., Cox, A. V., Craig, L, E., Smith, A. G., and Smith, D. G., eds., 1990, A geologic time scale 1989: Cambridge, United Kingdom, Cambridge University Press, 263 p.

Harland, W. B., Hambrey, M. J., and Waddams, P., 1993, Vendian geology of Svalbard: Oslo, Norsk Polarinstitutt, Skrifter Nr. 193, 150 p.

Harper, G. D., and Link, P. K., 1985, Geochemistry of rift-related metabasalts associated with Late Proterozoic diamictites, northern Utah and southeastern Idaho: Geology, v. 14, p. 864–867.

Harrison, C. G. A., 1990, Long-term eustasy and epeirogeny in continents, *in* Sea-level changes, Studies in geophysics: Washington, D.C., National Academy Press, p. 141–158.

Haug, G. H., and Tiedemann, R., 1998, Effect of the formation of the Isthmus of Panama on Atlantic Ocean thermohaline circulation: Nature, v. 393, p. 673–676.

Haughton, S. H., 1963, The stratigraphic history of Africa, south of the Sahara: New York, Hafner Publishing Company, 365 p.

Haughton, S. H., 1969, Geological history of southern Africa: Cape Town, Geological Society of South Africa, 535 p.

Hay, W. W., 1996, Tectonics and climate: Geologische Rundschau, v. 85, p. 409–437.

Hays, J. D., and Pitman, W. C., III, 1973, Lithospheric plate motion, sea level changes and climatic and ecological consequences: Nature, v. 256, p. 18–22.

Hays, J. D., Imbrie, J., and Shackleton, N. J., 1976, Variations in the Earth's orbit: Science, v. 194, p. 1121–1132.

Heaman, L. M., and Grotzinger, J. P., 1992, 1.08 Ga diabase sills in the Pahrump Group, California: Implications for development of the Cordilleran miogeocline: Geology, v. 20, p. 637–640.

Heckel, P. H., 1986, Sea-level curve for Pennsylvanian eustatic marine transgressive-regressive depositional cycles along midcontinent outcrop belt, North America: Geology, v. 14, p. 330–334.

Heckel, P. H., 1995, Glacial-eustatic base-level–climatic model for late middle to late Pennsylvanian coal-bed formation in the Appalachian basin: Journal of Sedimentary Research, v. B65, p. 348–356.

Heckel, P. H., and Witzke, B. J., 1979, Devonian world palaeogeography determined from distribution of carbonates and related palaeoclimatic indicators, *in* House, M. R., Scrutton, C. T., and Bassett, M. G., eds., The Devonian System: London, The Palaeontological Association, Special Papers in Palaeontology 23, p. 99–123.

Hefferan, K. P., Karson, J. A., and Saquaque, A., 1992, Proterozoic collisional basins in a Pan-African suture zone, Anti-Atlas Mountains, Morocco: Precambrian Research, v. 54, p. 295–319.

Heinrich, H., 1988, Origin and consequences of cyclic ice rafting in the Northeast Atlantic Ocean during the past 130,000 years: Quaternary Research, v. 29, p. 142–152.

Helwig, J., 1972, Stratigraphy, sedimentation, paleogeography, and paleoclimates of Carboniferous ("Gondwana") and Permian of Bolivia: American Association of Petroleum Geologists Bulletin, v. 56, p. 1008–1020.

Henry, G., Stanistreet, I. G., and Maiden, K. J., 1986, Preliminary results of a sedimentological study of the Chuos formation in the central zone of the Damara orogen: Evidence for mass flow processes and glacial activity: Communications of Geological Survey of SW Africa/Namibia, v. 2,

p. 75–92.

Herbert, T. D., and Fischer, A. G., 1986, Milankovitch climatic origin of mid-Cretaceous black shale rhythms in central Italy: Nature, v. 321, p. 739–743.

Heredia, S., and Beresi, M., 1995, Ordovician events and sea level changes on the western margin of Gondwana: The Argentine Precordillera, *in* Cooper, J. D., Drose, M. L., and Finney, S. C., eds., Ordovician Odyssey: Short papers for the seventh international symposium on the Ordovician System, Las Vegas, Nevada, USA: Fullerton, California, Pacific Section, Society for Sedimentary Geology (SEPM), p. 315–318.

Hettich, M., 1977, A glaciação proterozóica no centro-norte de Minas Gerais: Revista Basileira de Geociências, v. 7, p. 87–101.

Hoefs, J., 1987, Stable isotope geochemistry: Verlin, Springer-Verlag, 3rd edition, 241 p.

Hoffman, P. F., 1988, United plates of America, the birth of a craton: Early Proterozoic assembly and growth of Laurentia: Annual Review of Earth and Planetary Sciences, v. 16, p. 543–603.

Hoffman, P. F., 1989a, Speculations on Laurentia's first gigayear (2.0 to 1.0 Ga): Geology, v. 17, p. 135–138.

Hoffman, P. F., 1989b, Precambrian geology and tectonic history of North America, *in* Bally, A. W., and Palmer, A. R., eds., The Geology of North America— An Overview: Boulder, Colorado, Geological Society of America, The Geology of North America, v. A, p. 447–512.

Hoffman, P. F., 1991, Did the breakout of Laurentia turn Gondwanaland inside-out?: Science, v. 252, p. 1409–1412.

Hoffman, P. F., 1992, Relative timing of Rodinia breakup and Gondwanaland assembly: critical test of Laurentiacentric models for the Neoproterozoic supercontinent: EOS (American Geophysical Union Transactions), v. 73, p. 364.

Hoffman, P. F., Kaufman, A. J., and Halverson, G. P., 1998a, Comings and goings of global glaciations on a Neoproterozoic tropical platform in Namibia: GSA Today, v. 8, p. 1–9.

Hoffman, P. F., Kaufman, A. J., Halverson, G. P., and Schrag, D. P., 1998b, A Neoproterozoic snowball Earth: Science, v. 281, p. 1342–1346.

Hoffman, P. F., Kaufman, A. J., Halverson, G. P., and Schrag, D. P., 1999, A Neoproterozoic snowball Earth: Response: Science, v. 282, p. 1645–1646.

Hofmann, C., Courtillot, V., Féraud, G., Rochette, P., Yirgus, G., Ketefo, E., and Pik, R., 1997, Timing of the Ethiopian flood basalt event and implications for plume birth and global change: Nature, v. 389, p. 838–841.

Hofmann, H. J., 1987, Precambrian biostratigraphy: Geoscience Canada, v. 14, p. 135–154.

Hofmann, H. J., Narbonne, G. M., and Aitken, J. D., 1990, Ediacaran remains from intertillite beds in northeastern Canada: Geology, v. 18, p. 1199–1202.

Holland, H. D., 1978, The chemistry of the atmosphere and ocean: New York, Wiley, 351 p.

Holland, H. D., 1984, The chemical evolution of the atmosphere and oceans: Princeton, New Jersey, Princeton University Press, 582 p.

Holland, H. D., and Beukes, N. J., 1990, A paleoweathering profile from Griqualand West, South Africa: Evidence for a dramatic rise in atmospheric oxygen between 2.2 and 1.9 BYBP: American Journal of Science, v. 290-A, p. 1–34.

Holland, H. D., and Petersen, U., 1995, Living dangerously: Princeton, New Jersey, Princeton University Press, 490 p.

Holland, H. D., Lazar, B., and McCaffey, M., 1986, Evolution of the atmosphere and the oceans: Nature, v. 320, p. 27–33.

Hollin, J. T., 1969, Ice-sheet surges and the geological record: Canadian Journal of Earth Sciences, v. 6, p. 903–910.

Holmes, C. D., 1941, Till fabric: Geological Society of America Bulletin, v. 52, p. 1301–1354.

Holser, W. T., Schidlowski, M., Mackenzie, F. T., and Maynard, J. B., 1988, Biogeochemical cycles of carbon and sulfur, *in* Gregor, C. B., Garrels, R. M., Mackenzie, F. T., and Maynard, J. B., eds., Chemical cycles in the evolution of the earth: New York, John Wiley and Sons, p. 105–173.

Holtedahl, O., 1918, Bidrag til Finmarkens geologi: Norges Geologiske Undersøkelse, v. 11, p. 1–314.

Horodyski, R. J., 1993, Precambrian paleontology of the western conterminous

United States and northwestern Mexico, *in* Link, P. K., Christie-Blick, N., Devlin, W. J., Elston, D. P., Hirodyski, R. J., Levy, M., Miller, J. M. G., Pearson, R. C., Prave, A., Stewart, J. H., Winston, M., Wright, L. A., and Wrucke, C. T., 1993, Middle and Late Proterozoic stratified rocks of the western U. S. Cordillera, Colorado Plateau, and Basin and Range province, *in* Reed, J. C., Jr., Bickford, M. E., Houston, R. S., Link, P. K., Rankin, D. W., Sims, P. K., and Van Schmus, W. R., eds., Precambrian: Conterminous U.S.: Boulder, Colorado, Geological Society of America, The Geology of North America, v. C-2, p. 558–565.

Horodyski, R. J., and Knauth, L. P., 1994, Life on land in the Precambrian: Science, v. 263, p. 494–498.

Horz, F., Gall, H., Huttner, R., and Oberbeck, V. R., 1977, Shallow drilling in the "Bunte Breccia" impact deposits, Ries Crater, Germany, *in* Roddy, D. J., Pepin, R. O., and Merrill, R. B., eds., Impact and explosion cratering: New York, Pergamon, p. 425–448.

Horz, F., Ostertag, G. R., and Rainey, D. A., 1983, Bunte breccia of the Ries; continuous deposits of large impact craters: Reviews of Geophysics and Space Physics, v. 21, p. 1667–1725.

Houck, K. J., 1997, Effects of sedimentation, tectonics, and glacio-eustasy on depositional sequences, Pennsylvanian Minturn Formation, north-central Colorado: American Association of Petroleum Geologists Bulletin, v. 81, p. 1510–1533.

House, M. R., 1985a, A new approach to an absolute timescale from measurements of orbital cycles and sedimentary microrhythms: Nature, v. 315, p. 721–725.

House, M. R., 1985b, Correlation of mid-Palaeozoic ammonoid evolutionary events with global sedimentary perturbations: Nature, v. 313, p. 17–22.

House, M. R., 1989, Ammonoid extinction events: Royal Society, London, Philosophical Transactions, Series B, v. 325, p. 307–326.

House, M. R., and Gale, A. S., eds., 1995, Orbital forcing timescales and cyclostratigraphy: Geological Society of London Special Publication 85, 204 p.

Houston, R. S., 1993, Late Archean and Early Proterozoic geology of southeastern Wyoming, *in* Snoke, A. W., Steidtmann, J. R., and Roberts, S. M., eds., Geology of Wyoming: Geological Survey of Wyoming Memoir No. 5, p. 78–116.

Houston, R. S., and Karlstrom, K. E., 1993, Proterozoic rocks, *in* Houston, R. S., ed., The Wyoming province, Chapter 3, *in* Reed, J. C., Bickford, M. E., Houston, R. S., Link, P. K., Rankin, D. W., Sims, P. K., and Van Schmus, W. R., Precambrian: Conterminous U.S., The Geology of North America: Boulder, Colorado, Geological Society of America, v. C-2, p. 152–170.

Houston, R. S. Lanthier, L. R., Karlstrom, K. E., and Sylvester, G., 1981, Early Proterozoic diamictites of southern Wyoming, *in* Hambrey, M. J., and Harland, W. B., eds., Earth's pre-Pleistocene glacial record: Cambridge, United Kingdom, Cambridge University Press, p. 795–799.

Houston, R. S., Karlstrom, K. E., Graff, P. J., and Flurkey, A. J., 1992, New stratigraphic subdivisions and redefinitions of subdivisions of Late Archean and Early Proterozoic metasedimentary and metavolcanic rocks of the Sierra Madre and Medicine Bow Mountains, southern Wyoming: U.S. Geological Survey Professional Paper 1520, 50 p.

Howarth, R. J., 1971, The Port Askaig Tillite succession (Dalradian) of County Donegal: Royal Academy of Ireland Proceedings, v. 71, Section B, p. 1–35.

Howarth, R. J., Kilburn, C., and Leake, B. E., 1966, The boulder bed succession at Glencolumbkille, County Donegal: Royal Irish Academy Proceedings, v. 65, Section B, p. 117–156.

Hsü, K. J., 1985, Swiss lakes as a geological laboratory, Part II: Varves: Naturwissenschaften, v. 72, p. 365–371.

Hsü, K. J., 1995, The geology of Switzerland: Princeton, New Jersey, Princeton University Press, 250 p.

Hsü, K. J., Ryan, W. B. F., and Cita, M. B., 1973, Late Miocene desiccation of the Mediterranean: Nature, v. 242, p. 240–244.

Hubbert, M. K., 1967, Critique of the Principle of Uniformity, *in* Albritton, C. C., Jr., Uniformity and Simplicity: Geological Society of America Special Paper 89, p. 3–33.

Hudson, J. D., 1989, Palaeoatmospheres in the Phanerozoic: Geological Society of London Journal, v. 146, p. 27–33.

Hudson, J. D., and Anderson, T. F., 1989, Ocean temperatures and isotopic compositions through time: Royal Society of Edinburgh Transactions: Earth Sciences, v. 80, p. 183–192.

Huggett, R. J., 1991, Climate, Earth processes and Earth history: Berlin, Springer-Verlag, 281 p.

Hughes, D. W., 1996, Dust from beyond the Solar System: Nature, v. 380, p. 283.

Hunt, A. G., and Malin, P. E., 1998, Possible triggering of Heinrich events by ice-load–induced earthquakes: Nature, v. 393, p. 155–158.

Hurley, N. F., and Van der Voo, R., 1990, Magnetostratigraphy, Late Devonian iridium anomaly and impact hypotheses: Geology, v. 18, p. 291–294.

Idnurm, M., and Giddings, J. W., 1988, Australian Precambrian polar wander: A review: Precambrian Research, v. 40/41, p. 61–88.

Idnurm, M., and Giddings, J. W., 1995, Paleoproterozoic-Neoproterozoic North America-Australia link: New evidence from paleomagnetism: Geology, v. 23, p. 149–152.

Ilyin, A. V., 1990, Proterozoic supercontinent, its latest Precambrian rifting, breakup, dispersal into smaller continents, and subsidence of their margins: Evidence from Asia: Geology, v. 18, p. 1231–1234.

Imbrie, J., and Imbrie, J. Z., 1980, Modeling the climate response to orbital variations: Science, v. 207, p. 943–953.

Ireland, T. R., Flöttmann, T., Fanning, C. M., Gibson, G. M., and Preiss, W. V., 1998, Development of the early Paleozoic Pacific margin of Gondwana from detrital-zircon ages across the Delamerian orogen: Geology, v. 26, p. 243–246.

Irving, E., 1983, Fragmentation and assembly of the continents, mid-Carboniferous to present: Geophysical Surveys, v. 5, p. 299–333.

Isbell, J. L., Seegers, G. M., and Gelhar, G. A., 1997, Upper Paleozoic glacial and postglacial deposits, central Transantarctic Mountains, Antarctica, *in*, Martini, I. P., ed., Late Glacial and Postglacial Environmental Changes: Quaternary, Carboniferous-Permian, and Proterozoic: New York, Oxford University Press, p. 230–242.

Isotta, C. A. L., Rocha-Campos, A. C., and Yoshida, R., 1969, Striated pavement of the upper Precambrian glaciation in Brazil: Nature, v. 222, p. 466–468.

Isozaki, Y., 1997, Permo-Triassic boundary superanoxia and stratified superocean: Records from lost deep sea: Science, v. 276, p. 235–238.

Ito, T., Masuda, K., Hamano, Y., and Matsui, T., 1995, Climate friction: A possible cause for secular drift of Earth's obliquity: Journal of Geophysical Research, v. 100, p. 15147–15161.

Jackson, J. B. C., Jung, P., Coates, A. G., and Collins, L. S., 1993, Diversity and extinction of tropical American mollusks and emergence of the Isthmus of Panama: Science, v. 260, p. 1624–1629.

Jenkins, G. S., and Frakes, L. A., 1998, GCM sensitivity test using increased rotation rate, reduced solar forcing and orography to examine low latitude glaciations in the Neoproterozoic: Geophysical Research Letters, v. 25, p. 3525–3528.

Jenkins, G. S., and Scotese, C. R., 1998, An early snowball Earth?: Discussion: Science, v. 282, p. 1644–1645.

Jenkins, R. J. F., 1990, The Adelaide fold belt: Tectonic reappraisal, *in* Jago, J. B., and Moore, P. S., eds., The evolution of a late Precambrian–early Palaeozoic rift complex: the Adelaide geosyncline: Geological Society of Australia Special Publication 16, p. 396–420.

Jensen, P. A., and Wulff-Pedersen, E., 1996, Glacial or non-glacial origin for the Bigganjargga tillite, Finnmark, northern Norway: Geological Magazine, v. 133, p. 137–145.

Jensen, P. A., and Wulff-Pedersen, E., 1997, Glacial or non-glacial origin for the Bigganjargga tillite, Finnmark, northern Norway: Reply: Geological Magazine, v. 134, p. 874–876.

Joachimski, M. M., and Buggisch, W., 1993, Anoxic events in the later Frasnian—Causes of the Frasnian-Famennian faunal crisis?: Geology, v. 21, p. 675–678.

Johnson, B. K., 1957, Geology of part of the Manly Peak Quadrangle, southern Panamint Range, California: University of California Publications in Geological Sciences, v. 30, p. 353–424.

Johnson, J. G., and Potter, E. C., 1976, Late Ordovician—Early Silurian glaciation

and the Ordovician-Silurian boundary in northern Canadian Cordillera: Comment: Geology, v. 4, p. 795.

Johnson, J. G., Klapper, G., and Sandberg, C. A., 1985, Devonian eustatic fluctuations in Euramerica: Geological Society of America Bulletin, v. 96, p. 567–587.

Johnson, M. E., 1996, Stable cratonic sequences and a standard for Silurian eustasy, *in* Witzke, B. J., Ludvigson, G. A., and Dag, J., eds., Paleozoic Sequence Stratigraphy: Views from the North American Craton: Boulder, Colorado, Geological Society of America Special Paper 306, p. 203–211.

Jopling, A. V., and McDonald B. C., eds., 1975, Glaciofluvial and glaciolacustrine sedimentation: Society of Economic Paleontologists and Mineralogists Special Publication 23, 320 p.

Karhu, J. A., and Holland, H. D., 1996, Carbon isotopes and the rise of atmospheric oxygen: Geology, v. 24, p. 867–870.

Kasting, J. F., 1987, Theoretical constraints on oxygen and carbon dioxide concentrations in the Precambrian atmosphere: Precambrian Research, v. 34, p. 205–229.

Kasting, J. F., 1992a, Proterozoic climates: the effect of changing atmospheric carbon dioxide concentrations, *in* Schopf, J. W., and Klein, C., eds., The Proterozoic biosphere: New York, Cambridge University Press, p. 165–168.

Kasting, J. F., 1992b, Paradox lost and paradox found: Nature, v. 355, p. 676–677.

Kasting, J. F., 1993, Earth's early atmosphere: Science, v. 259, p. 920–926.

Kasting, J. F., and Ackerman, T. P., 1986, Climatic consequences of very high carbon dioxide levels in the Earth's early atmosphere: Science, v. 234, p. 1383–1385.

Kasting, J. F., Holland, H. D., and Kump, L. R., 1992, Atmospheric evolution: the rise of oxygen, *in* Schopf, J. W., and Klein, C., eds., The Proterozoic biosphere: New York, Cambridge University Press, p. 159–163.

Kauffman, S., 1995, At home in the universe: New York, Oxford University Press, 321 p.

Kaufman, A. J., and Knoll, A. M., 1995, Neoproterozoic variations in the C-isotopic composition of seawater: Stratigraphic and biogeochemical implications: Precambrian Research, v. 73, p. 27–49.

Kaufman, A. J., Knoll, A. H., and Awramik, S. M., 1992, Biostratigraphic and chemostratigraphic correlation of Neoproterozoic sedimentary successions: Upper Tindir Group, northwestern Canada, as a test case: Geology, v. 20, p. 181–185.

Kaufman, A. J., Knoll, A. H., and Narbonne, G. M., 1997, Isotopes, ice ages, and terminal Proterozoic earth history: Proceedings National Academy of Sciences, v. 94, p. 6600–6605.

Keller, C. K., and Wood, B. D., 1993, Possibility of chemical weathering before the advent of vascular land plants: Nature, v. 364, p. 223–225.

Keller, M., 1995, Continental slope deposits in the Argentine Precordillera: sediments and geotectonic significance, *in* Cooper, J. D., Droser, M. L., and Finney, S. C., eds., Ordovician odyssey: Short papers for the seventh international symposium on the Ordovician System, Las Vegas, Nevada: Fullerton, California, Pacific Section, Society for Sedimentary Geology (SEPM), p. 211–215.

Kelley, P. M., and Wigley, T. M. L., 1992, Solar cycle length, greenhouse forcing and global climate: Nature, v. 360, p. 328–330.

Kemp, A. E. S., ed., 1996, Palaeoclimatology and palaeoceanography from laminated sediments: Geological Society of London Special Paper 116, 272 p.

Kemper, E., 1983, Uber Kalt- und Warmzeiten der Unterkreide: Zittcliana, v. 10, p. 359–369.

Kennedy, M. J., 1996, Stratigraphy and isotopic geochemistry of Australian Neoproterozoic postglacial cap dolostones: Deglaciation, δ^{13} excursions, and carbonate precipitation: Journal of Sedimentary Research, v. 66, p. 1050–1064.

Kennedy, M. J., Runnegar, B., Prave, A. R., Hoffman, K.-H., and Arthur, M. A., 1998, Two or four Neoproterozoic glaciations?: Geology, v. 36, p. 1059–1063.

Kennett, J. P., and Thunell, R. C., 1975, Global increase in Quaternary explosive volcanism: Science, v. 187, p. 497–503.

Kenrick, P., and Crane, P. R., 1997, The origin and early evolution of plants on land: Nature, v. 389, p. 33–39.

Kent, D. E., and Van der Voo, R., 1990, Palaeozoic palaeogeography from palaeomagnetism of the Atlantic-bordering continents, *in* McKerrow, W. S., and Scotese, C. R., eds., Palaeozoic Palaeogeography and Biogeography: Geological Society of London Memoir No. 12, p. 49–56.

Kiehl, J. T., 1994, Clouds and their effects on the climate system: Physics Today, v. 47, p. 36–42.

Kilburn, C., Pitcher, W. S., and Shackleton, R. M., 1965, The stratigraphy and origin of the Port Askaig Boulder Bed Series (Dalradian): Geological Journal, v. 4, p. 343–360.

Kimura, H., Matsumoto, R., Kakuwa, Y., Hamdi, B., and Zibaseresht, H., 1997, The Vendian-Cambrian δ¹³C record, north Iran: Evidence for overturning the ocean before the Cambrian explosion: Earth and Planetary Science Letters, v. 147, p. E1–E7.

Kirschvink, J. L., 1992, Late Proterozoic low-latitude global glaciation: the snowball Earth, *and* A paleogeographic model for Vendian and Cambrian time, *in* Schopf, J. W., and Klein, C., eds., The Proterozoic biosphere: New York, Cambridge University Press, p. 51–52; 569–581.

Kirschvink, J. L., Ripperdan, R. L., and Evans, D. A., 1997, Evidence for a large-scale reorganization of early Cambrian continental masses by inertial interchange true polar wander: Science, v. 277, p. 541–545.

Klein, C., and Beukes, N. J., 1992, Time distribution, stratigraphy, and sedimentologic setting, and geochemistry of Precambrian iron formations: Palaeogeography, Palaeoclimatology, Palaeoecology, v. 98, p. 139–141.

Klein, C., and Beukes, N. J., 1993, Sedimentology and geochemistry of the glaciogenic Late Proterozoic Rapitan iron-formation in Canada: Economic Geology, v. 88, p. 542–565.

Klein, G. D., 1990, Pennsylvanian time scales and cycle periods: Geology, v. 18, p. 455–457.

Klein, G. D., 1991, Pennsylvanian time scales and cycle periods: Reply: Geology, v. 19, p. 405–406.

Klein, G. D., 1992, Climatic and tectonic sea-level gauge for Midcontinent Pennsylvanian cyclothems: Geology, v. 20, p. 363–366.

Klein, G. D. ed., 1994, Pangea: Paleoclimate, tectonics, and sedimentation during accretion, zenith, and breakup of a supercontinent: Geological Society of America Special Paper 288, 295 p.

Klein, G. D., and Beauchamp, B., 1994, Introduction: Project Pangea and workshop recommendations, *in* Klein G. D., ed., Pangea: Paleoclimate, Tectonics, and Sedimentation During Accretion, Zenith, and Breakup of a Supercontinent: Boulder, Colorado, Geological Society of America Special Paper 288, p. 1–12.

Klein, G. D., and Kupperman, J. B., 1992, Pennsylvanian cyclothems: Methods of distinguishing tectonically induced changes in sea level from climatically induced changes: Geological Society of America Bulletin, v. 104, p. 166–175.

Klein, G. D., and Willard, D. A., 1989, Origin of the Pennsylvanian coal-bearing cyclothems of North America: Geology, v. 17, p. 152–155.

Klitzsch, E., 1983, Paleozoic formations and a Carboniferous glaciation from the Gilf Kebir–Abu Ras area in southwestern Egypt: Journal of African Earth Sciences, v. 1, p. 17–19.

Knauth, L. P., and Epstein, S., 1976, Hydrogen and oxygen isotope ratios in nodular and bedded cherts: Geochimica et Cosmochimica Acta, v. 40, p. 1095–1108.

Knoll, A. H., and Walter, M. R., 1992, Latest Proterozoic stratigraphy and Earth history: Nature, v. 356, p. 673–678.

Knoll, A. H., and Walter, M. R., eds., 1995, Neoproterozoic stratigraphy and Earth history: Precambrian Research (Special volume), v. 73, 298 p.

Knoll, A. H., Hayes, J. M., Kaufman, A. J., Swett, K., and Lambert, I. B., 1986, Secular variations in carbon isotope ratios from Upper Proterozoic successions of Svalbard and East Greenland: Nature, v. 321, p. 812–838.

Knoll, A. H., Bambach, R. K., Canfield, D. E., and Grotzinger, J. P., 1996, Comparative Earth history and late Permian mass extinction: Science, v. 273, p. 452–457.

Korsch, R. J., and Lindsay, J. F., 1989, Relationships between deformation and basin evolution in the intracratonic Amadeus basin, central Australia: Tectonophysics, v. 158, p. 5–22.

Kortenkamp, S. J., and Dermott, S. F., 1998, A 100,000-year periodicity in the accretion rate of interplanetary dust: Science, v. 280, p. 874–876.

Koster, E. H., and Steel, R. J., eds., 1984, Sedimentology of gravels and conglomerates: Canadian Society of Petroleum Geologists Memoir 10, 441 p.

Kreuser, T., Wopfner, H., Kaaya, C. Z., Markwort, S., Semkiwa, P., and Aslanidis, P., 1990, Depositional evolution of Permo-Triassic Karoo basins in Tanzania with reference to their economic potential: Journal of African Earth Sciences, v. 10, p. 151–167.

Krinsley, D. H., and Doornkamp, J. C., 1973, Atlas of quartz sand surface textures: Cambridge, United Kingdom, Cambridge University Press, 91 p.

Krogh, T. E., Davis, D. W., and Corfu, F., 1984, Precise U-Pb zircon and baddeleyite ages for the Sudbury area, *in* Pye, E. G., Naldrett, A. J., and Giblin, P. E., eds., The geology and ore deposits of the Sudbury structure: Toronto, Ontario Geological Survey Special Volume 1, p. 431–436.

Kröner, A., 1977, Non-synchroneity of Late Precambrian glaciations in Africa: Journal of Geology, v. 85, p. 289–300.

Kröner, A., McWilliams, M. O., Germs, G. J. B., Reid, A. B., and Schalk, K. E. L., 1980, Paleomagnetism of Late Precambrian to Early Paleozoic mixtite-bearing formations in Namibia (South West Africa): the Nama Group and Blaubeker Formation: American Journal of Science, v. 280, p. 942–968.

Kruck, W., and Thiele, J., 1983, Late Palaeozoic glacial deposits in the Yemen Republic: Geologisches Jahrbuch, Reihe B, v. 46, p. 3–29.

Kukla, G., Berger, A., Lotti, R., and Brown, J., 1981, Orbital signature of interglacials: Nature, v. 290, p. 295–300.

Kukla, P. A., and Stanistreet, I. G., 1991, Record of the Damaran Khomas Hochland accretionary prism in central Namibia: Refutation of an "ensialic" origin of a Late Proterozoic orogenic belt: Geology, v. 18, p. 473–476.

Kump, L. R., 1989, Alternative modeling approaches to the geochemical cycles of carbon, sulfur, and strontium isotopes: American Journal of Science, v. 289, p. 390–410.

Kump, L. R., Griggs, M. T., Arthur, M. A., Patzkowsky, M. E., and Sheehan, P. M., 1995, Hivnantian glaciation and the carbon cycle, *in* Cooper, J. D., Droser, M. L., and Finney, S. C., eds., Ordovician Odyssey: Short papers for the seventh international symposium on the Ordovician System, Las Vegas, Nevada: Fullerton, California, Pacific Section, Society for Sedimentary Geology (SEPM), p. 299–302.

Kumpulainen, R., and Nystuen, J. P., 1985, Late Proterozoic basin evolution and sedimentation in the westernmost part of Baltoscandia, *in* Gee, D. G., and Sturt, B. A., eds., The Caledonide Orogen—Scandinavia and related areas: New York, John Wiley and Sons, p. 213–232.

Kutzbach, J. E., Prell, W. L., and Ruddiman, W. F., 1993, Sensitivity of Eurasian climate to surface uplift of the Tibetan Plateau: Journal of Geology, v. 101, p. 177–190.

Laird, M. G., 1972, Sedimentation of the ?Late Precambrian Raggo Group, Varanger Peninsula, *and* The stratigraphy and sedimentology of the Laksefjord Group, Finnmark: Norges Geologiske Undersøkelse, v. 278, p. 1–40.

Langenheim, R. L., Jr., 1991, Pennsylvanian time scales and cycle periods: Comment: Geology, v. 19, p. 405.

Larsen, H. C., Saunders, A. D., Clift, P. D., Beget, J., Wei, W., Spezzaferri S., and ODP Leg 152 Scientific Party, 1994, Seven million years of glaciation in Greenland: Science, v. 264, p. 952–955.

Larson, R. L., 1991, Latest pulse of Earth: Evidence for a mid-Cretaceous superplume: Geology, v. 19, p. 547–550.

Laskar, J., and Robutel, P., 1993, The chaotic obliquity of the planets: Nature, v. 361, p. 608–612.

Leggett, J. K., McKerrow, W. S., Cocks, L. R. M., and Rickards, R. B., 1981, Periodicity in the early Palaeozoic marine realm: Geological Society of London Journal, v. 138, p. 167–176.

Legrand, P., 1995, Evidence and concerns with regard to the Late Ordovician glaciation in North Africa, *in* Cooper, J. D., Droser, M. L., and Finney, S. C., eds.: Ordovician Odyssey: Short papers for the seventh international symposium on the Ordovician System, Las Vegas, Nevada: Fullerton, California, Pacific Section, Society for Sedimentary Geology (SEPM), p. 165–169.

Lenz, A. C., 1976, Late Ordovician–Early Silurian glaciation and the Ordovician-Silurian boundary in the northern Canadian Cordillera: Geology, v. 4,

p. 313–317.

Lenz, A. C., 1976, Late Ordovician–Early Silurian glaciation and the Ordovician-Silurian boundary in the northern Canadian Cordillera: Replies: Geology, v. 4, p. 795–797.

Lenz, A. C., 1982, Ordovician to Devonian sea-level changes in western and northern Canada: Canadian Journal of Earth Sciences, v. 19, p. 1919–1932.

Levell, B. K., Braakman, J. H., and Rutten, K. W., 1988, Oil-bearing sediments of Gondwana glaciation in Oman: American Association of Petroleum Geologists Bulletin, v. 72, p. 775–796.

Levy, M., and Christie-Blick, N., 1989, Pre-Mesozoic palinspastic reconstruction of the eastern Great Basin (western United States): Science, v. 145, p. 1454–1464.

Levy, M., and Christie-Blick, N., 1991, Tectonic subsidence of the early Paleozoic passive continental margin in eastern California and southern Nevada: Geological Society of America Bulletin, v. 103, p. 1590–1606.

Lewis, C. F. M., and Keen, J. J., 1990, Constraints to development, *in* Keen, J. J., and Williams, G. L., eds., Geology of the continental margin of eastern Canada: Ottawa, Canada, Geological Survey of Canada, v. 2; p. 785–793.

Li, Z. X., and Powell, C. McA., 1993, Late Proterozoic to early Paleozoic paleomagnetism and the formation of Gondwanaland, *in* Findlay, R. H., Unrug, R., Banks, M. R., and Veevers, J. J., eds., Gondwana Eight; Assembly, evolution and dispersal: Rotterdam, Netherlands, Balkema, p. 9–21.

Li, Z. X., Zhang, L., and Powell, C. McA., 1995, South China in Rodinia: part of the missing link between Australia-East Antarctica and Laurentia?: Geology, v. 23, p. 407–410.

Li, Z. X., Zhang, L., and Powell, C. McA., 1996, Positions of the East Asian cratons in the Neoproterozoic supercontinent of Rodinia: Australian Journal of Earth Sciences, v. 43, p. 593–604.

Lin, Jin-lu, Fuller, M., and Zhang, Wen-you, 1985, Preliminary Phanerozoic polar wander paths for the North and South China blocks: Nature, v. 313, p. 444–449.

Lindsay, J. F., 1970, Depositional environment of Paleozoic glacial rocks in the central Transantarctic Mountains: Geological Society of America Bulletin, v. 84, p. 1149–1172.

Lindsay, J. F., 1989, Depositional controls on glacial facies associations in a basinal setting, Late Proterozoic, Amadeus Basin, central Australia: Palaeogeography, Palaeoclimatology, Palaeoecology, v. 73, p. 205–232.

Lindsay, J. F., 1997, Permian postglacial environments of the Australian plate, *in* Martini, I. P., ed., Late Glacial and Postglacial Environmental Changes: New York, Oxford University Press, p. 213–229.

Lindsay, J. F., Korsch, R. J., and Wilford, J., 1987, Timing the breakup of a Proterozoic supercontinent: Evidence from Australian intracratonic basins: Geology, v. 15, p. 1061–1064.

Lindsay, J. F., Brasier, M. D., Shields, G., Bat-Ireedui, Y. A., 1996, Glacial facies associations in a Neoproterozoic back-arc setting, Sarkhan basin, western Mongolia: Geological Magazine, v. 133, p. 391–402.

Lindsey, D. A., 1969, Glacigenic rocks in the Early Proterozoic Chibougamau Formation of northern Quebec: Geological Society of America Bulletin, v. 80, p. 1685–1702.

Lindsey, D. A., 1971, Glacial marine sediments in the Precambrian Gowganda Formation of Whitefish Falls, Ontario, Canada: Palaeogeography, Palaeoclimatology, Palaeoecology, v. 9, p. 7–25.

Link, P. K., 1983, Glacial and tectonically influenced sedimentation in the upper Proterozoic Pocatello Formation, southeastern Idaho, *in* Miller, D. M., Todd, V. R., and Howard, K. A., eds., Tectonic and stratigraphic studies in the eastern Great Basin: Geological Society of America Memoir 157, p. 165–182.

Link, P. K., and Gostin, V. A., 1981, Facies and paleogeography of Sturtian glacial strata (Late Precambrian), South Australia: American Journal of Science, v. 281, p. 353–374.

Link, P. K., Christie-Blick, N., Devlin, W. J., Elston, D. P., Horodyski, R. J., Levy, M., Miller, J. M. G., Pearson, R. C., Prave, A., Stewart, J. H., Winston, D., Wright, L. A., and Wrucke, C. T., 1993, Middle and Late Proterozoic stratified rocks of the western U.S. Cordillera, Colorado Plateau, and Basin and Range province, *in* Reed, J. C., Jr., Bickford, M. E., Houston,

R. S., Link, P. K., Rankin, D. W., Simms, P. K., and Van Schmus, W. R., eds., Precambrian: Conterminous U.S.: Boulder, Colorado, Geological Society of America, The Geology of North America, v. C-2, p. 463–595.

Link, P. K., Christie-Blick, N., Stewart, J. H., Miller, J. M. G., Devlin, W. J., and Levy, M. E., 1993b, Late Proterozoic strata of the United States cordillera, *in* Reed, J. C., Jr., 1993, Introduction, *in* Reed, J. C., Jr., Bickford, M. E., Houston, R. S., Link, P. K., Rankin, D. W., Sims, P. K., and Van Schmus, W. R., eds., Precambrian: Conterminous U.S.: Boulder, Colorado, Geological Society of America, The Geology of North America, v. C-2, p. 536–558.

Link, P. K., Miller, J. M. G., and Christie-Blick, N., 1994, Glacial-marine facies in a continental rift environment: Neoproterozoic rocks of the western United States Cordillera, *in* Deynoux, M., Miller, J. M. G., Domack, E. W., Eyles, N., Fairchild, I. J., and Young, G. M., eds., Earth's glacial record: Cambridge, United Kingdom, Cambridge University Press, p. 29–46.

Linnemann, U.-G., 1995, The Neoproterozoic terranes of Saxony (Germany): Precambrian Research, v. 73, p. 235–250.

Liu Hung-yun, Sha Ching-an, and Hu Shi-ling, 1973, The Sinian System in southern China: Scientia Sinica, v. 16, p. 266–278.

Loboziak, S., Streel, M., Caputo, M. V., and de Melo, J. H. G., 1992, Middle Devonian to lower Carboniferous miospore stratigraphy in the central Parnaíba Basin (Brazil): Annales de la Société Géologique de Belgique, v. 115, p. 215–226.

Logan, G. A., Hayes, J. M., Hieshima, G. B., and Summons, R. E., 1995, Terminal Proterozoic reorganization of biogeochemical cycles: Nature, v. 376, p. 53–56.

Lohmann, H. H., 1965, Paläozoische Vereisungen in Bolivien: Geologische Rundschau, v. 54, p. 161–165.

Long, D. G. F., 1973, The stratigraphy and sedimentology of the Chibougamau Formation (Aphebian), Chibougamau, Quebec: Canadian Journal of Earth Sciences, v. 11, p. 1236–1252.

Long, D. G. F., 1981, Glacigenic rocks in the Early Proterozoic Chibougamau Formation of northern Quebec, *in* Hambrey, M. J., and Harland, W. B., eds., Earth's pre-Pleistocene glacial record: Cambridge, United Kingdom, Cambridge University Press, p. 817–820.

Long, D. G. F., 1991, A non-glacial origin for the Ordovician (Middle Caradocian) Cosquer Formation, Veryarc'h, Crozon Peninsula, Brittany, France: Geological Journal, v. 26, p. 279–293.

Lonsdale, P., 1989, Geology and tectonic history of the Gulf of California, *in* Winterer, E. L., Hussong, D. M., and Decker, R. W., eds., The Eastern Pacific Ocean and Hawaii: Boulder, Colorado, Geological Society of America, The Geology of North America, v. N, p. 499–521.

López Gamundi, O. R., 1987, Depositional models for the glaciomarine sequences of Andean Late Paleozoic basins of Argentina: Sedimentary Geology, v. 52, p. 109–126.

López Gamundi, O. R., 1989, Postglacial transgressions in Late Paleozoic basins of western Argentina: A record of glacioeustatic sea level rise: Palaeogeography, Palaeoclimatology, Palaeoecology, v. 71, p. 257–270.

López Gamundi, O. R., 1991, Thin-bedded diamictites in the glaciomarine Hoyada Verde Formation (Carboniferous), Calingasta-Uspallata basin, western Argentina: A discussion on the emplacement conditions of subaqueous cohesive debris flows: Sedimentary Geology, v. 73, p. 247–256.

López-Gamundi, O. R., 1997, Glacial-postglacial transition in the late Paleozoic basins of southern South America, *in* Martini, I. P., ed., Late Glacial and Postglacial Environmental Changes: New York, Oxford University Press, p. 147–168.

López-Gamundi, O. R., and Rossello, E. A., 1998, Basin fill evolution and paleotectonic patterns along the Samfrau geosyncline: the Sauce Grande basin–Ventana foldbelt (Argentina) and Karoo basin–Cape foldbelt (South Africa) revisited: Geologische Rundschau, v. 86, p. 819–834.

López-Gamundi, O. R., Espejo, I. S., Conaghan, P. J., and Powell, C. McA., with a contribution by Veevers, J. J., 1994, Southern South America, *in* Veevers, J. J., and Powell, C. McA., eds., Permian-Triassic Pangean basins and foldbelts along the Panthalassan margin of Gondwanaland: Geological Society of America Memoir 184, p. 281–329.

Lorius, C., Jouzel, J., Raynaud, D., Hansen, J., and Le Treut, H. E., 1990, The ice-core record: climate sensitivity and future greenhouse warming: Nature, v. 347, p. 139–145.

Lovelock, J. E., and Kump, L. R., 1994, Failure of climate regulation in a geophysiological model: Nature, v. 369, p. 732–734.

Lowe, D. R., 1979, Sediment gravity flows: their classification and some problems of application to natural flows and deposits, in Doyle, L. J., and Pilkey, O. H., eds., Geology of continental slopes: Society of Economic Paleontologists and Mineralogists Special Publication 27, p. 75–84.

Lowe, D. R., Byerly, G. R., Asaro, F., and Kyte, F. J., 1989, Geological and geochemical record of 3400-million-year-old terrestrial meteorite impacts: Science, v. 245, p. 959–962.

Lowe, D. R., Grotzinger, J. P., and Ingersoll, R. V., 1992, Introduction; Proterozoic sedimentary basins; the Proterozoic sedimentary record; Major events in the geological development of the Precambrian Earth; Summary and conclusions, in Schopf, J. W., and Klein, C., eds., The Proterozoic Biosphere: New York, Cambridge University Press, p. 45–50, 53–57, 67–77.

Lu Songnian and Qu Lesheng, 1987, Characteristics of the Sinian glaciogenic rocks of the Shennongjia Region, Hubei Province, China: Precambrian Research, v. 36, p. 127–142.

Lu Songnian, Ma Guogan, Gao Zhenjia, and Lin Weixing, 1985, Sinian ice ages and glacial sedimentary facies–areas in China: Precambrian Research, v. 29, p. 53–63.

Maack, R., 1957, Uber Vereisungsperioden und Vereisungsspuren in Brasilien: Geologische Rundschau, v. 45, p. 547–595.

MacNiocaill, C., van der Pluijm, B. A., and Van der Voo, R., 1997, Ordovician paleogeography and evolution of the Iapetus ocean: Geology, v. 25, p. 159–162.

Magaritz, M., Krishnamurthy, R. V., and Holser, W. T., 1992, Parallel trends in organic and inorganic carbon isotopes across the Permian/Triassic boundary: American Journal of Science, v. 292, p. 727–739.

Mahaney, W. C., 1995, Glacial crushing, weathering and diagenetic histories of quartz grains inferred from electron microscopy, in Menzies, J., ed., Modern Glacial Environments: Oxford, United Kingdom, Butterworth-Heinemann, p. 487–506.

Manby, G. M., and Hambrey, M. J., 1989, The structural setting of the Late Proterozoic tillites of East Greenland, in Gayer, R. A., ed., The Caledonide geology of Scandinavia: London, Graham and Trotman, p. 299–312.

Marmo. J. S., and Ojakangas, R. W., 1984, Lower Proterozoic glaciogenic deposits, eastern Finland: Geological Society of America Bulletin, v. 95, p. 1055–1062.

Marshall, J. D., and Middleton, P. D., 1990, Changes in marine isotopic composition and the Late Ordovician glaciation: Geological Society of London Journal, v. 147, p. 1–4.

Marshall, J. E. A., 1994, The Falkland Islands: A key element in Gondwana paleogeography: Tectonics, v. 13, p. 499–514.

Martin, D. McB., Stanistreet, I. G., and Camden-Smith, P. M., 1989, The interaction between tectonics and mudflow deposits within the Main Conglomerate Formation in 2.8–2.7 Ga Witwatersrand Basin: Precambrian Research, v. 44, p. 19–38.

Martin, H., 1961, The hypothesis of continental drift in the light of recent advances of geological knowledge in Brazil and Southwest Africa: A. L. du Toit Memorial Lecture No. 7: Geological Society of South Africa Transactions, Annexure, v. 44, p. 1–47.

Martin, H., 1981a, The Late Palaeozoic Dwyka Group of the South Kalahari Basin in Namibia and Botswana, and the subglacial valleys of the Kaokoveld in Namibia, in Hambrey, M. J., and Harland, W. B., eds., Earth's pre-Pleistocene glacial record: Cambridge, United Kingdom, Cambridge University Press, p. 61–66.

Martin, H., 1981b, The Late Palaeozoic Dwyka Group of the Karasburg Basin, Namibia, in Hambrey, M. J., and Harland, W. B., eds., Earth's pre-Pleistocene glacial record: Cambridge, United Kingdom, Cambridge University Press, p. 67–70.

Martin, H., 1981c, The Palaeozoic Gondwana glaciation: Geologische Rundschau, v. 70, p. 480–496.

Martin, H., Porada, H., and Walliser, O. H., 1985, Mixtite deposits of the Damara sequence, Namibia, problems of interpretation: Palaeogeography, Palaeoclimatology, Palaeoecology, v. 51, p. 159–196.

Matsch, C. L., and Ojakangas, R. W., 1992, Stratigraphy and sedimentology of the Whiteout Conglomerate—A late Paleozoic glacigenic sequence in the Ellsworth Mountains, West Antarctica, in Webers, G. F., Craddock, C., and Splettstoesser, J. F., eds., Geology of the Ellsworth Mountains, Antarctica: Geological Society of America Memoir 170, p. 37–62.

Max, M. D., 1981, Dalradian tillite of northwestern Ireland, in Hambrey, M. J., and Harland, W. B., eds., Earth's pre-Pleistocene glacial record: Cambridge, United Kingdom, Cambridge University Press, p. 640–642.

Maynard, J. R., and Leeder, M. R., 1992, On the periodicity and magnitude of Late Carboniferous glacio-eustatic sea level changes: Geological Society of London Journal, v. 149, p. 303–311.

McCann, A. M., and Kennedy, M. J., 1974, A probable glaciomarine deposit of Late Ordovician–Early Silurian age from the north central Newfoundland Appalachian belt: Geological Magazine, v. 11, p. 549–564.

McClure, H. A., 1978, Early Paleozoic glaciation in Arabia: Palaeogeography, Palaeoclimatology, Palaeoecology, v. 25, p. 315–326.

McClure, H. A., 1980, Permian-Carboniferous glaciation in the Arabian Peninsula: Geological Society of America Bulletin, Part I, v. 91, p. 207–712.

McClure, H. A., and Young, G. M., 1981, Late Palaeozoic glaciation in the Arabian Peninsula, in Hambrey, M. J., and Harland, W. B., eds., Earth's pre-Pleistocene Glacial Record: Cambridge, United Kingdom, Cambridge University Press, p. 275–277.

McCormick, M. P., Thomason, L. W., and Trepte, C. R., 1995, Atmospheric effects of the Mt Pinatubo eruption: Nature, v. 373, p. 399–404.

McCrea, W. H., 1975, Ice ages and the galaxy: Nature, v. 255, p. 607–609.

McCrea, W. H., 1981, Long time-scale fluctuations in the evolution of the Earth: Royal Society of London Proceedings, Series A, v. 375, p. 1–41.

McElhinny, M. W., and Embleton, B. J. J., 1974, Australian palaeomagnetism and the Phanerozoic plate tectonics of eastern Gondwanaland: Tectonophysics, v. 22, p. 1–29.

McElhinny, M. W., Giddings, J. W., and Embleton, B. J. J., 1974, Palaeomagnetic results and late Precambrian glaciations: Nature, v. 248, p. 557–561.

McGhee, G. R., Jr., 1996, The Late Devonian mass extinction: The Frasnian/Famennian crisis: New York, Columbia University Press, 303 p.

McGillivray, J. G., and Husseini, M. I., 1992, The Paleozoic petroleum geology of central Arabia: American Association of Petroleum Geologists Bulletin, v. 76, p. 1473–1490.

McKelvey, V. E., 1978, Phosphates in sediments, in Fairbridge, R. W., and Bourgeois, J., eds., The encyclopedia of sedimentology: Stroudsburg, Pennsylvania, Dowden, Hutchinson and Ross, Inc., p. 574–579.

McKerrow, W. S., 1979, Ordovician and Silurian changes in sea-level: Geological Society of London Journal, v. 136, p. 137–145.

McKerrow, W. S., and Scotese, C. R., eds., 1990, Palaeozoic palaeogeography and biogeography: Geological Society of London Memoir 12, 435 p.

McKinnon, W. B., 1997, Extreme cratering: Science, v. 276, p. 1346–1348.

McLaren, D. J., 1970, Time, life, and boundaries: Journal of Paleontology, v. 44, p. 801–815.

McLaren, D. J., 1983, Bolides and stratigraphy: Geological Society of America Bulletin, v. 94, p. 313–324.

McLaren, D. J., and Goodfellow, W. D., 1990, Geological and biological consequences of giant impacts: Annual Review of Earth and Planetary Sciences, v. 18, p. 123–171.

McMenamin, M. A. S., and McMenamin, D. L. S., 1990, The emergence of animals: the Cambrian breakthrough: New York, Columbia University Press, 217 p.

McPhie, J., 1987, Andean analogue for Late Carboniferous volcanic arc and arc flank environments of the western New England orogen, New South Wales, Australia: Tectonophysics, v. 138, p. 269–288.

McWilliams, M. O., 1981, Palaeomagnetism and Precambrian tectonic evolution of Gondwana, in Kröner, A., ed., Precambrian plate tectonics: Amsterdam, Elsevier, p. 649–687.

McWilliams, M. O., and Kröner, A., 1981, Palaeomagnetism and tectonic evolution of the Pan-African Damara belt, southern Africa: Journal of Geophysical Research, v. 86, p. 5147–5162.

McWilliams, M. O., and McElhinny, M. W., 1980, Late Precambrian palaeomagnetism of Australia, the Adelaide geosyncline: Journal of Geology, v. 88, p. 1–26.

Meert, J. G., and van der Voo, R., 1994, The Neoproterozoic (1000–540 Ma) glacial intervals: No more snowball earth?: Earth and Planetary Science Letters, v. 123, p. 1–13.

Meert, J. G., and van der Voo, R., 1995, The Neoproterozoic (1100–540 Ma) glacial intervals: No more snowball Earth? Reply: Earth and Planetary Science Letters, v. 131, p. 123–125.

Meert, J. G., and van der Voo, R., 1996, Paleomagnetic and ^{40}Ar/^{39}Ar studies of the Sinyai dolerite, Kenya: Implications for Gondwana assembly: Journal of Geology, v. 104, p. 131–142.

Meert, J. G., Van der Voo, R., and Ayub, S., 1995, Paleomagnetic investigation of the Neoproterozic Gagwe lavas and Mbozi complex, Tanzania and the assembly of Gondwana: Precambrian Research, v. 74, p. 225–244.

Menard, H. W., 1973, Epeirogeny and plate tectonics: EOS (American Geophysical Union Transactions), v. 54, p. 1244–1255.

Menzies, J., ed., 1995, Modern glacial environments: Processes, dynamics, sediments: Oxford, United Kingdom, Butterworth-Heinemann, Ltd., 621 p.

Menzies, J., ed., 1996, Past glacial environments: Sediments, forms and techniques: Oxford, United Kingdom, Butterworth-Heinemann, Ltd., 598 p.

Metcalfe, I., 1988, Origin and assembly of south-east Asian continental terranes, *in* Audley-Charles, M. G., and Hallam, A., eds., Gondwana and Tethys: Geological Society of London Special Publication No. 37, p. 101–118.

Mezger, K., van der Pluijm, B. A., Essene, E. J., and Halliday, A. N., 1991, Synorogenic collapse: A perspective from the middle crust, the Proterozoic Grenville orogen: Science, v. 254, p. 695–700.

Miall, A. D., 1983, Glaciomarine sedimentation in the Gowganda Formation (Huronian), northern Ontario: Journal of Sedimentary Petrology, v. 53, p. 477–491.

Miall, A. D., 1985, Sedimentation on an early Proterozoic continental margin under glacial influence: the Gowganda Formation (Huronian), Elliot Lake area, Ontario, Canada: Sedimentology, v. 32, p. 763–788.

Miall, A. D., 1992, Exxon global cycle chart: An event for every occasion?: Geology, v. 20, p. 787–790.

Miller, J. M. G., 1985, Glacial and syntectonic sedimentation: The upper Proterozoic Kingston Peak Formation, southern Panamint Range, eastern California: Geological Society of America Bulletin, v. 96, p. 1537–1553.

Miller, J. M. G., 1987, Paleotectonic and stratigraphic implications of the Kingston Peak–Noonday contact in the Panamint Range, eastern California: Journal of Geology, v. 95, p. 75–85.

Miller, J. M. G., 1989, Glacial advance and retreat sequences in a Permo-Carboniferous section, central Transantarctic Mountains: Sedimentology, v. 36, p. 419–430.

Miller, J. M. G., 1994, The Neoproterozoic Konnarock Formation, southwestern Virginia, USA: glaciolacustrine facies in a continental rift, *in* Deynoux, M., Miller, J. M. G., Domack, E. W., Eyles, N., Fairchild, I. J., and Young, G. M., eds., Earth's glacial record: Cambridge, United Kingdom, Cambridge University Press, p. 47–59.

Miller, J. M. G., 1996, Glacial sediments, *in* Reading, H. G., Sedimentary environments: Processes, facies and stratigraphy: Oxford, United Kingdom, Blackwell Science, p. 454–484.

Miller, J. M. G., and Waugh, B., 1991, Permo-Carboniferous glacial sedimentation in the central Transantarctic Mountains and its palaeotectonic significance, *in* Thompson, M. R. A., Crame, J. A., and Thompson, J. W., eds., Geological evolution of Antarctica: Cambridge, United Kingdom, Cambridge University Press, p. 205–208.

Miller, J. M. G., Troxel, B., and Wright, L. A., 1988, Stratigraphy and paleogeography of the Proterozoic Kingston Peak Formation, Death Valley region, eastern California, *in* Gregory, J. L., and Baldwin, E. J., eds., Geology of the Death Valley region, Santa Ana, California, South Coast Geological Society Annual Field Trip Guidebook No. 16: p. 118–142.

Miller, K. G., Fairbanks, R. G., and Mountain, G. S., 1987, Tertiary oxygen isotope synthesis, sea level history and continental margin erosion: Paleoceanography, v. 1, p. 1–20.

Miller, K. B., McCahon, T. J., and West, R. R., 1996a, Lower Permian (Wolfcampian) Paleosol-bearing cycles of the U.S. midcontinent: Evidence of climatic cyclicity: Journal of Sedimentary Research, v. 66, p. 71–84.

Miller, K. G., Mountain, G. S., the Leg 150 Shipboard Party, and members of the New Jersey coastal plain drilling project, 1996b, Drilling and dating New Jersey Oligocene-Miocene sequences: Ice volume, global sea level, and Exxon records: Science, v. 271, p. 1092–1095.

Miller, R. H., 1976, Comments on Lenz, A. C., 1976, Late Ordovician–Early Silurian glaciation and the Ordovician-Silurian boundary in the northern Canadian Cordillera: Geology, v. 4, p. 796.

Minnis, P., Harrison, E. F., Stowe, L. L., Gibson, G. G., Denn, F. M., Doelling, D. R., and Smith, W. L., Jr., 1993, Radiative climate forcing by the Mount Pinatubo eruption: Science, v. 259, p. 1411–1415.

Mitchell, C., Taylor, G. K., Cox, K. G., and Shaw, J., 1986, Are the Falkland Islands a rotated microplate?: Nature, v. 319, p. 131–134.

Molnar, P., and England, P., 1990, Late Cenozoic uplift of mountain ranges and global climate change: chicken or egg?: Nature, v. 346, p. 29–34.

Molnia, B. F., ed., 1983a, Glacial-marine sedimentation: New York, Plenum Press, 844 p.

Molnia, B. F., 1983b, Subarctic glacial-marine sedimentation: A model, *in* Molnia, B. F., ed., Glacial-marine sedimentation: New York, Plenum Press, p. 95–144.

Moncrieff, A. C. M., 1989, The Tillite Group and related rocks of East Greenland: implications for Late Proterozoic palaeogeography, *in* Gayer, R. A., ed., The Caledonide geology of Scandinavia: London, Graham and Trotman, p. 285–297.

Moncrieff, A. C. M., and Hambrey, M. J., 1988, Late Precambrian glacially related grooved and striated surfaces in the Tillite Group of central East Greenland: Palaeogeography, Palaeoclimatology, Palaeoecology, v. 65, p. 183–200.

Moncrieff, A. C. M., and Hambrey, M. J., 1990, Marginal-marine glacial sedimentation in the late Precambrian succession of East Greenland: *in* Dowdeswell, J. A., and Scourse, J. D., eds., Glacimarine environments: Processes and sediments: Geological Society of London Special Publication No. 53, p. 387–410.

Mooney, W. D., and Braile, L. W., 1989, The seismic structure of the continental crust and upper mantle of North America, *in* Bally, A. W., and Palmer, A. R., eds., The Geology of North America—An overview: Boulder, Colorado, Geological Society of America, The Geology of North America, v. A, p. 39–52.

Moore, R. C., 1964, Paleoecological aspects of Kansas Pennsylvanian and Permian cyclothems, *in* Merriam, D. F., ed., Symposium on cyclic sedimentation: Kansas Geological Survey Bulletin 169, p. 287–380.

Moore, T. C., Jr., Pisias, N. G., and Keigwin, L. D., Jr., 1982, Cenozoic variability of oxygen isotopes in benthic Foraminifera, *in* Climate in Earth History, Studies in Geophysics: Washington, D.C., National Academy Press, p. 172–182.

Moores, E. M., 1991, Southwest U.S.–East Antarctica (SWEAT) connection: A hypothesis: Geology, v. 19, p. 425–428.

Moores, E. M., 1993, Neoproterozoic oceanic crustal thinning, emergence of continents, and origin of the Phanerozoic ecosystem: A model: Geology, v. 21, p. 5–8.

Moores, E. M., 1994, Neoproterozoic oceanic crustal thinning, emergence of continents, and origin of the Phanerozoic ecosystem: A model: Reply: Geology, v. 22, p. 88.

Morgan, J., Warner, M., and the Chicxulub working group, Brittan, J., Buffler, R., Camargo, A., Christeson, G., Denton, P., Hildebrand, A., Hobbs, R., Macintyre, H., Mackenzie, G., Maguire, P., Marin, L., Nakamura, Y., Pilkingtron, M., Sharpton, V., Snyder, D., Suarez, G., and Trejo, A., 1997, Size and morphology of the Chicxulub impact crater: Nature, v. 390, p. 472–476.

Morgan, W. J., 1981, Hotspot tracks and the opening of the Atlantic and Indian oceans, *in* Emiliani, C., ed., The Sea, v. 7: New York, J. Wiley and Sons,

p. 443–487.

Mörner, N.-A., 1978, Varves and varved clays, *in* Fairbridge, R. W., and Bourgeois, J., Encyclopedia of Sedimentology: Stroudsburg, Pennsylvania, Dowden, Hutchinson and Ross, Inc., p. 841–843.

Morris, W. A., 1977, Paleolatitude of glaciogenic upper Precambrian Rapitan Group and the use of tillites as chronostratigraphic marker horizons: Geology, v. 5, p. 85–88.

Morris, W. A., and Aitken, J. D., 1982, Paleomagnetism of the Little Dal lavas, Mackenzie Mountains, Northwest Territories, Canada: Canadian Journal of Earth Sciences, v. 19, p. 2020–2027.

Moura, P., 1938, Geologia do Baixo Amazonas: Rio de Janeiro, Serviço Geológico e Mineralógico, Bullitin 91, 94 p.

Mukasa, S. B., and Henry, D. J., 1990, The San Nicholas Batholith of coastal Peru: Early continental arc or continental rift magmatism: Geological Society of London Journal, v. 147, p. 27–39.

Muller, R. A., and MacDonald, G. J., 1995, Glacial cycles and orbital inclinations: Nature, v. 377, p. 108.

Muller, R. A., and MacDonald, G. J., 1997, Simultaneous presence of orbital inclination and eccentricity in proxy climate record from Ocean Drilling Program Site 806: Geology, v. 25, p. 3–6.

Murphy, D. M., Anderson, J. R., Quinn, P. K., McInnes, L. M., Brechtel, F. J., Kreidenweis, S. M., Middlebrook, A. M., Pósfal, M., Thomson, D. S., and Buseck, P. R., 1998, Influence of sea-salt on aerosol radiative properties in the Southern Ocean marine boundary layer: Nature, v. 392, p. 62–64.

Mustard, P. S., 1991, Normal faulting and alluvial-fan deposition, basal Windermere tectonic assemblage, Yukon, Canada: Geological Society of America Bulletin, v. 103, p. 1346–1364.

Mustard, P. S., and Donaldson, J. A., 1987, Early Proterozoic ice-proximal glaciomarine deposition: The lower Gowganda Formation at Cobalt, Ontario, Canada: Geological Society of America Bulletin, v. 98, p. 373–387.

Mu Yongji, 1981, Luoquan tillites of the Sinian System in China, *in* Hambrey, M. J., and Harland, W. B., eds., Earth's pre-Pleistocene glacial record: Cambridge, United Kingdom, Cambridge University Press, p. 402–413.

Myers, J. S., Shaw, R. D., and Tyler, I. M., 1996, Tectonic evolution of Proterozoic Australia: Tectonics, v. 15, p. 1431–1446.

Myers, R. E., McCarthy, T. S., and Stanistreet, I. G., 1990, A tectono-sedimentary reconstruction of the development and evolution of the Witwatersrand Basin, with particular emphasis on the Central Rand Group: South African Journal of Geology, v. 93, p. 180–201.

Myrow, P. M., 1995, Neoproterozoic rocks of the Newfoundland Avalon zone: Precambrian Research, v. 73, p. 123–136.

Nance, R. D., Worsley, T. R., and Moody, J. G., 1986, Post-Archean biogeochemical cycles and long-term episodicity in tectonic processes: Geology, v. 14, p. 514–518.

Narbonne, G. M., and Aitken, J. D., 1995, Neoproterozoic of the Mackenzie Mountains, northwestern Canada: Precambrian Research, v. 73, p. 101–121.

Narbonne, G. M., Kaufman, A. J., and Knoll, A. H., 1994, Integrated chemostratigraphy and biostratigraphy of the Windermere Supergroup, northwestern Canada: Implications for Neoproterozoic correlations and the early evolution of animals: Geological Society of America Bulletin, v. 106, p. 1281–1292.

NAS-NRC Panel, 1982, Solar variability, weather, and climate [Studies in geophysics]: Washington, D.C., National Academy Press, 106 p.

NAS-NRC Panel, 1983, Changing climate: Report of the carbon dioxide assessment committee: Washington, National Academy Press, 496 p.

NAS-NRC Panel, 1990, Sea-level change [Studies in Geophysics]: Washington, D.C., National Academy Press, 234 p.

Negrutsa, T. F., and Negrutsa, V. Z., 1981, Early Proterozoic Lammos tilloids of the Kola Peninsula; Early Proterozoic Yanis-Yarvi tilloids, south Karelia; Early Proterozoic Sarioili tilloids in the eastern part of the Baltic Shield; Archaean Pebozero tilloids of Karelia; Archaean tilloids of the Tikshozero Group in the eastern part of the Baltic Shield, U.S.S.R., *in* Hambrey, M. J., and Harland, W. B., eds., Earth's pre-Pleistocene glacial record: Cambridge, United Kingdom, Cambridge University Press, p. 678–690.

Nesbitt, H. W., and Young, G. M., 1982, Early Proterozoic climates and plate

motions inferred from major element chemistry of lutites: Nature, v. 299, p. 715–717.

Nesbitt, H. W., and Young, G. M., 1996, Petrogenesis of sediments in the absence of chemical weathering: effects of abrasion and sorting on bulk composition and mineralogy: Sedimentology, v. 43, p. 341–358.

Nesbitt, H. W., Young, G. M., McLennan S. M., and Keays, R. R., 1996, Effects of chemical weathering on the petrogenesis of siliciclastic sediments, with implications for provenance studies: Journal of Geology, v. 104, p. 525–542.

Nie, S., Rowley, D. B., and Ziegler, A. M., 1990, Constraints on the location of the Asian microcontinents in Palaeo-Tethys during the Late Palaeozoic, *in* McKerrow, W. S., and Scotese C. R., eds., Palaeozoic palaeogeography and biogeography, Geological Society of London Memoir 12, p. 397–410.

Niklas, K. J., Tiffney, B. H., and Knoll, A. H., 1985, Patterns in vascular land plant diversification: an analysis at the species level, *in* Valentine, J. W., ed., Phanerozoic diversity patterns: Profiles in macroevolution: Princeton, New Jersey, Princeton University Press, p. 97–128.

Nunes, P. D., and Tilton, G. R., 1971, Uranium-lead ages of minerals from the Stillwater Complex, Montana: Geological Society of America Bulletin, v. 82, p. 2231–2250.

Nystuen, J. P., 1983, Nappe and thrust structure in the Sparagmite Region, southern Norway: Norges Geologiske Undersøkelse, v. 380, p. 67–83.

Nystuen, J. P., 1985, Facies and preservation of glaciogenic sequences from the Varanger Ice Age in Scandinavia and other parts of the North Atlantic region: Palaeogeography, Palaeoclimatology, Palaeoecology, v. 51, p. 209–229.

Nystuen, J. P., and Sæther, T., 1979, Clast studies in the Late Precambrian Moelv Tillite and Osdal Conglomerate, Sparagmite Region, south Norway: Norges Geologiske Undersøkelse, v. l59, p. 239–251.

Oberbeck, V. R., Marshall, J. R., and Aggarwal, H., 1993, Impacts, tillites, and the breakup of Gondwanaland: Journal of Geology, v. 101, p. 1–19.

O'Driscoll, E. S. T., 1986, Observations of the lineament-ore relation: Royal Society of London Philosophical Transactions, Series A, v. 317, p. 195–218.

Oerlemans, J., 1993, Evaluating the role of climate cooling in iceberg production and the Heinrich events: Nature, v. 364, p. 783–786.

Ohmoto, H., 1996, Evidence in pre-2.2 Ga paleosols for the early evolution of atmospheric oxygen and terrestrial biota: Geology, v. 24, p. 1135–1138.

Ojakangas, R. W., 1985, Evidence for Early Proterozoic glaciation: the dropstone unit-diamictite association: Geological Survey of Finland Bulletin 351, p. 51–72.

Ojakangas, R. W., 1988, Glaciation: An uncommon "mega-event" as a key to intracontinental and intercontinental correlation of Early Proterozoic basin fill, North American and Baltic Cratons, *in* Kleinspehn, K. L., and Paola, C., eds., New perspectives in basin analysis: New York, Springer-Verlag, p. 431–444.

Ojakangas, R. W., 1991, Regolith in early Proterozoic glacigene deposits: climate change due to plate motion or nontectonic causes?, *in* Report of Project 260, Earth's glacial record: International Geological Correlation Program, annual meeting and field trip, Bamko, Mali, January, 7-17, 1991, Abstract, p. 15.

Ojakangas, R. W., and Matsch, C. L., 1980, Upper Precambrian (Eocambrian) Mineral Fork Tillite of Utah: A continental glacial glaciomarine sequence: Geological Society of America Bulletin, Part 1, v. 91, p. 495–501.

Ojakangas, R. W., and Matsch, C. L., 1981, The Late Palaeozoic Whiteout Conglomerate: A glacial and glacialmarine sequence in the Ellsworth Mountains, West Antarctica, *in* Hambrey, M. J., and Harland, W. B., eds., Earth's pre-Pleistocene glacial record: Cambridge, United Kingdom, Cambridge University Press, p. 241–244.

Ojakangas, R. W., and Matsch, C. L., 1982, Upper Precambrian (Eocambrian) Mineral Fork Tillite of Utah: A continental glacial glaciomarine sequence: Reply: Geological Society of America Bulletin, Part 1, v. 93, p. 186–187.

Oldow, J. S., Bally, A. W., Avé Lallemant, H. G., and Leeman, W. P., 1989, Phanerozoic evolution of the North American Cordillera; United States and Canada, *in* Bally, A. W., and Palmer, A. R., eds., The Geology of North America—An overview: Boulder, Colorado, Geological Society of America, The Geology of North America, v. A, p. 139–232.

Oliver, G. J. H., Johnson, S. P., Williams, I. S., and Herd, D. A, 1998, Relict 1.4 Ga oceanic crust in the Zambezi Valley, northern Zimbabwe: evidence for Mesoproterozoic supercontinental fragmentation: Geology, v. 26, p. 571–573.

Oliver, J., 1996, Shocks and rocks: Seismology in the plate tectonics revolution: Washington, D.C., American Geophysical Union, 130 p.

Olsen, P. E., 1986, A 40-million year lake record of early Mesozoic orbital climatic forcing: Nature, v. 234, p. 842–848.

Olsen, P. E., and Kent, D. V., 1996, Milankovitch climate forcing in tropics of Pangaea during the Late Triassic: Palaeogeography, Palaeoclimatology, Palaeoecology, v. 122, p. 1–26.

Opdyke, N. D., Roberts, J., Claoué-Long, J., and Irving, E., personal communication about magnetic stratigraphy of the type Kiaman and the age of the base of the Permo-Carboniferous reversed superchron.

Ovenshine, A. T., 1970, Observations of iceberg rafting in Glacier Bay, Alaska, and the identification of ancient ice rafted deposits.

Page, N. J., 1977, Stillwater Complex, Montana; Rock succession, metamorphism, and structure of the complex and adjacent rocks: U.S. Geological Survey Professional Paper 999, 79 p.

Page, N. J., 1981, The Precambrian diamictite below the base of the Stillwater Complex, Montana, in Hambrey, M. J., and Harland, W. B., eds., Earth's pre-Pleistocene glacial record: Cambridge, United Kingdom, Cambridge University Press, p. 821–825.

Page, N. J., and Koski, R. A., 1973, A Precambrian diamictite below the base of the Stillwater Complex, southwestern Montana: U.S. Geological Survey Journal of Research, v. 1, p. 403–414.

Palmer, A. R., 1983, The decade of North American geology; 1983 geologic time scale: Geology, v. 11, p. 503–504.

Panahi, A., and Young, G. M., 1997, A geochemical investigation into the provenance of the Neoproterozoic Port Askaig Tillite, Dalradian Supergroup, Western Scotland: Precambrian Research, v. 85, p. 81–96.

Paris, F., Elaouad-Debbaj, Z., Jaglin, J. C., Massa, D., and Oulebsir, L., 1995, Chitinozoans and late Ordovician glacial events on Gondwana, in Cooper J. D., Droser, M. L., and Finney, S. C., eds., Ordovician odyssey: Short papers for the seventh international symposium on the Ordovician System, Las Vegas, Nevada: Fullerton, California, Pacific Section, Society for Sedimentary Geology (SEPM), p. 171–176.

Parish, J. T., 1993, Climate of the supercontinent Pangea: Journal of Geology, v. 101, p. 215–233.

Parish, J. T., Demko, T. M., and Tanck, G. S., 1993, Sedimentary palaeoclimatic indicators: What they are and what they tell us: Royal Society of London Philosophical Transactions, Series A, v. 344, p. 21–25.

Park, J. K., 1997, Paleomagnetic evidence for low-latitude glaciation during deposition of the Neoproterozoic Rapitan Group, Mackenzie Mountains, N.W.T., Canada: Canadian Journal of Earth Sciences, v. 34, p. 34–39.

Park, R. G., 1997, Early Precambrian plate tectonics: South African Journal of Geology, v. 100, p. 23–35.

Patterson, J. G., and Heaman, L. M., 1991, New geochronologic limits on the depositional age of the Hurwitz Group, Trans-Hudson hinterland, Canada: Geology, v. 19, p. 1137–1140.

Paytan, A., Kastner, M., Campbell, D., and Thiemens, M. H., 1998, Sulfur isotopic composition of Cenozoic seawater sulfate: Science, v. 282, p. 1459–1462.

Pederson, T. F., and Calvert, S. E., 1990, Anoxia vs. productivity: What controls the formation of organic-carbon–rich sediments and sedimentary rocks?: American Association of Petroleum Geologists Bulletin, v. 74, p. 454–466.

Pedrosa-Soares, A. C., Vidal, P., Leonardos, O. H., and De Brito-Nevis, B. B., 1998, Neoproterozoic ocean remnants in eastern Brazil: Further evidence and refutation of an exclusively ensialic evolution for the Araçuaí–West Congo orogen: Geology, v. 26, p. 519–522.

Peixoto, J. P., and Oort, A. H., 1992, Physics of climate: New York, American Institute of Physics, 520 p.

Pekar, S., and Miller, K. G., 1996, New Jersey Oligocene "icehouse" sequences (ODP Leg 150X) correlated with global $\delta^{18}O$ and Exxon eustatic records: Geology, v. 14, p. 567–570.

Pesonen, L. J., Torsvik, T. H., Elming, S.-Å, and Bylund, G., 1989, Crustal evolution of Fennoscandia—palaeomagnetic constraints: Tectonophysics, v. 162, p. 27–49.

Peterman, Z. E., Hedge, C. E., and Tourtelot, H. A., 1970, Isotopic composition of strontium in sea water throughout Phanerozoic time: Geochimica et Cosmochimica Acta, v. 34, p. 105–120.

Pettijohn, F. J., 1943, Basal Huronian conglomerates of Menominee and Calumet districts, Michigan: Journal of Geology, v. 51, p. 387–397.

Pflug, R., and Schöll, W. U., 1975, Proterozoic glaciations in eastern Brazil: a review: Geologische Rundschau, v. 64, p. 287–299.

Pickerell, R. K., Pajari, G. E., and Currie, K. L., 1979, Evidence of Caradocian glaciation in the Davidsville group of northeastern Newfoundland: Geological Survey of Canada, Paper 79-1c, p. 67–72.

Pimentel, M. M., and Fuck, R. A., 1992, Neoproterozoic crustal accretion in central Brazil: Geology, v. 20, p. 375–379.

Piper, D. J. W., Mudie, P. J., Fader, G. B., Josenhaus, H. W., MacLean, B., and Vilks, G., 1990, Quaternary history, in Keen, M. J., and Williams, G. L., eds., Geology of the continental margin of eastern Canada: Ottawa, Canada, Geological Survey of Canada, v. 2, p. 475–607.

Piper, J. D. A., 1987, Palaeomagnetism and the continental crust: Milton Keynes, United Kingdom, Open University Press, and New York, John Wiley and Sons, Ltd., 434 p.

Pitman, W. C., III, 1978, Relationship between eustacy and stratigraphic sequences of passive margins: Geological Society of America Bulletin, v. 89, p. 1389–1403.

Playford, P. E., McLaren, D. J., Orth, C. J., Gilmore, J. S., and Goodfellow, W. D., 1984, Iridium anomaly in the Upper Devonian of the Canning Basin, Western Australia: Science, v. 226, p. 437–439.

Plumb, K. A., 1991, New Precambrian time scale: Episodes, v. 14, p. 139–140.

Plumb, K. A., 1993, Note on revised correlations of Neoproterozoic glacigene sequences of the Kimberley, northern Australia: Discussion note for 1993 field trip to Australia, working group on terminal Proterozoic system, International Union of Geological Sciences, p. 1–11.

Pollack, J. B., 1982, Solar, astronomical, and atmospheric effects on climate, in Climate in Earth History, Studies in Geophysics: Washington, D.C., National Academy Press, p. 68–76.

Powell, C. McA., 1993, Assembly of Gondwanaland—Open forum, in Findlay, R. H., Unrug, R., Banks, M. R., and Veevers, J. J., eds., Gondwana Eight, assembly, evolution and dispersal: Rotterdam, Netherlands, Balkema, p. 219–237.

Powell, C. McA., 1995, Are Neoproterozoic glacial deposits preserved on the margins of Laurentia related to the fragmentation of two supercontinents?: Comment: Geology, v. 23, p. 1053–1054.

Powell, C. McA., and Li, Z. X., 1994, Reconstruction of the Panthalassan margin of Gondwanaland, in Veevers, J. J., and Powell, C. McA., eds., Permian-Triassic Pangean basins and foldbelts along the Panthalassan margin of Gondwanaland: Geological Society of America Memoir 184, p. 5–9.

Powell, C. McA., and Veevers, J. J., 1987, Namurian uplift in Australia and South America triggered the main Gondwanan glaciation: Nature, v. 326, p. 177–179.

Powell, C. McA., Li, Z. X., McElhinny, M. W., Meert, J. G., and Park, J. K., 1993, Paleomagnetic constraints on the Neoproterozoic breakup of Rodinia and the mid-Cambrian formation of Gondwanaland: Geology, v. 21, p. 889–892.

Powell, C. McA., Preiss, W. V., Gatehouse, C. G., Krapez, B., and Li, Z. X., 1994, South Australian record of a Rodinian epicontinental basin and its mid-Neoproterozoic breakup (~700 Ma): Tectonophysics, v. 237, p. 113–140.

Powell, R. D., and Neumiller, C. M., 1986, Glacigenic evidence for Makganyene diamictite (2200 Ma) and low oceanic ^{18}O: International Association of Sedimentologists 12th Congress, Canberra, Australia, Abstracts, p. 246.

Prave, A. R., 1996, Tale of three cratons: Tectonostratigraphic anatomy of the Damara orogen in northwestern Namibia and the assembly of Gondwana: Geology, v. 24, p. 1115–1118.

Preiss, W. V., compiler, 1987, The Adelaide Geosyncline—late Proterozoic stratigraphy, sedimentation, palaeontology and tectonics: Geological Survey of South Australia Bulletin, v. 53, 438 p.

Preiss, W. V., 1990, A stratigraphic and tectonic overview of the Adelaide Geosyncline, South Australia, *in* Jago, J. B., and Moore, P. S., eds., The evolution of a late Precambrian–early Paleozoic rift complex: the Adelaide Geosyncline: Geological Society of Australia Special Publication 16, p. 1–33.

Pringle, I. R., 1973, Rb/Sr age determination on shales associated with the Varanger Ice Age: Geological Magazine, v. 109, p. 465–472.

Proust, J. N., and Deynoux, M., 1994, Marine to non-marine sequence architecture of an intracratonic glacially related basin: Late Proterozoic of the West African platform in western Mali, *in* Deynoux, M., Miller, J. M. G., Domack, E. W., Eyles, N., Fairchild, I. J., and Young, G. M., eds., Earth's glacial record: Cambridge, United Kingdom, Cambridge University Press, p. 121–145.

Puffett, W. P., 1969, The Reany Creek Formations, Marquette County, Michigan: U.S. Geological Survey Bulletin 1274-F, p. F1–F25.

Qi Rui Zhang, 1994, Environmental evolution during the early phase of Late Proterozoic glaciation, Hunan, China, *in* Deynoux, M., Miller, J. M. G., Domack, E. W., Eyles, N., Fairchild, I. J., and Young, G. M., eds., Earth's glacial record: Cambridge, United Kingdom, Cambridge University Press, p. 260–266.

Quilty, P. G., 1984, Phanerozoic climates and environments of Australia, *in* Veevers, J. J., ed., Phanerozoic Earth history of Australia: Oxford, United Kingdom, Clarendon Press, p. 48–57.

Radick, R. R., Lockwood, G. W., and Baliunas, S. L., 1990, Stellar activity and brightness variations: A glimpse at the Sun's history: Science, v. 247, p. 39–44.

Railsback, L. B., Ackerley, S. D., Anderson, T. F., and Cisne, J. L., 1990, Palaeontological and isotope evidence for warm saline deep waters in Ordovician oceans: Nature, v. 343, p. 156–159.

Rainbird, R. H., Jefferson, C. W., and Young, G. M., 1996, The early Neoproterozoic sedimentary Succession B of northwestern Laurentia: Correlations and paleogeographic significance: Geological Society of America Bulletin, v. 108, p. 454–470.

Ramos, V. A., Jordan, T. E., Allmendinger, R. W., Mpodozis, C., Kay, S. M., Cortés, J. M., and Palma, M. A., 1986, Paleozoic terranes of the central Argentine Chilean Andes: Tectonics, v. 5, p. 855–880.

Rampino, M. R., 1994, Tillites, diamictites, and ballistic ejecta of large impacts: Journal of Geology, v. 102, p. 439–456.

Rampino, M. R., and Self, S., 1992, Volcanic winter and accelerated glaciation following the Toba super-eruption: Nature, v. 359, p. 50–51.

Rampino, M. R., and Stothers, R. B., 1984, Geological rhythms and cometary impacts: Science, v. 226, p. 1427–1531.

Rampino, M. R., Self, S., and Stothers, R. B., 1988, Volcanic winters: Annual Review of Earth and Planetary Sciences: v. 16, p. 73–99.

Ramsay, A. C., 1855, On the occurrence of angular, subangular, polished, and striated fragments and boulders in the Permian breccia of Shropshire, Worcestershire, etc., and on the probable existence of glaciers and icebergs in the Permian Epoch: Geological Society of London Quarterly Journal, v. 11, p. 185–205.

Ramsay, A. C., 1880, On the recurrence of certain phenomena in geological time: Nature, v. 22, p. 383–390.

Rankin, D. W., 1975, The continental margin of eastern North America in the southern Appalachians: the opening and closing of the proto-Atlantic Ocean: American Journal of Science, v. 275A, p. 298–336.

Rankin, D. W., 1993, The volcanogenic Mount Rogers Formation and the overlying glaciogenic Konnarock Formation—two Late Proterozoic units in southwestern Virginia: U.S. Geological Survey Bulletin 2024, 26 p.

Rankin, D. W., Chiarenzelli, J. R., Drake, A. A., Jr., Goldsmith, R., Hall, L. M., Hinze, W. J., Isachsen, Y. W., Lidiak, E. G., McLelland, J., Mosher, S., Ratcliffe, N. M., Secor, D. T., Jr., and Whitney, P. R., 1993, Proterozoic rocks east and southeast of the Grenville front, *in* Reed, J. C., Jr., Bickford, M. E., Houston, R. S., Link, P. K., Rankin, D. W., Sims, P. K., and Van Schmus, W. R., eds., Precambrian: Conterminous U.S.: Boulder, Colorado, Geological Society of America, The Geology of North America, v. C-2,

p. 335–461.

Raymo, M. E., 1991, Geochemical evidence supporting T. C. Chamberlin's theory of glaciation: Geology, v. 19, p. 344–347.

Raymo, M. E., 1994a, The Himalayas, organic carbon burial, and climate in the Miocene: Paleoceanography, v. 9, p. 399–404.

Raymo, M. E., 1994b, The initiation of Northern Hemisphere glaciation: Annual Review of Earth and Planetary Sciences, v. 22, p. 353–383.

Raymo, M. E., 1997, Carbon cycle models: How strong are the constraints? *in* Ruddiman, W. F., ed., Tectonic uplift and climate change: New York, Plenum Press, p. 367–381.

Raymo, M. E., and Ruddiman, W. F., 1992, Tectonic forcing of late Cenozoic climate: Nature, v. 359, p. 117–122.

Raymo, M. E., and Ruddiman, W. F., 1994, Tectonic forcing of late Cenozoic climate: Reply: Nature, v. 361, p. 124.

Raymo, M. E., Ruddiman, W. F., and Froelich, P. N., 1988, Influence of late Cenozoic mountain building on ocean geochemical cycles: Geology, v. 16, p. 649–653.

Raymo, M. E., Ruddiman, W. F., Backman, J., Clement, B. M., and Martinson, D. G., 1989, Late Pliocene variations in northern hemisphere ice sheets and North Atlantic Deep Water circulation: Paleoceanography, v. 4, p. 413–446.

Reading, H. G., and Walker, R. G., 1966, Sedimentation of Eocambrian tillites and associated sediments in Finnmark, northern Norway: Palaeogeography, Palaeoclimatology, Palaeoecology, v. 2, p. 177–212.

Reed, J. C., Jr., Bickford, M. E., Houston, R. S., Link, P. K., Rankin, D. W., Sims, P. K., and Van Schmus, W. R., eds., 1993, Precambrian: Conterminous U.S.: Boulder, Colorado, Geological Society of America, The Geology of North America, v. C-2, 657 p.

Reid, P. C., and Tucker, M. E., 1972, Probable late Ordovician glacial marine sediments from northern Sierra Leone: Nature, v. 238, p. 38–40.

Reimold, W. U., Von Brunn, V., and Koeberl, C., 1997, Are diamictites impact ejecta?—No supporting evidence from South African Dwyka Group diamictites: Journal of Geology, v. 105, p. 517–530.

Renne, P. R., and Basu, A. R., 1991, Rapid eruption of the Siberian Traps flood basalts at the Permo-Triassic boundary: Science, v. 253, p. 176–179.

Renne, P. R., Ernesto, M., Pacca, I. G., Coe, R. S., Glen, J. M., Prévot, M., and Perrin, M., 1992, The age of Paraná flood volcanism, rifting of Gondwanaland, and the Jurassic-Cretaceous boundary: Science, v. 258, p. 975–979.

Renne, P. R., Zhang, Z., Richards, M. A., Black, M. T., and Basu, A. R., 1995, Synchrony and causal relations between Permian-Triassic boundary crises and Siberian flood volcanism: Science, v. 269, p. 1413–1416.

Retallack, G. J., 1995, Permian-Triassic life crisis on land: Science, v. 267, p. 77–79.

Retallack, G. J., 1997, Early forest soils and their role in Devonian global change: Science, v. 276, p. 583–585.

Retallack, G. J., Seyedolali, A., Krull, E. S., Holser, W. T., Ambers, C. P., and Kyte, F. T., 1998, Search for evidence of impact at the Permian-Triassic boundary in Antarctica and Australia: Geology, v. 26, p. 979–982.

Reusch, B. H., 1891, Sk:uringsmærker og moranegrums eftervist i Finmarker fra en periode meget neldre end "istiden": Norges Geologiske Undersøgelse, Årbok 1891, p. 78–85.

Rich, P. V., Rich, T. H. V., Wagstaff, B. E., McEwen, M. J., Douthitt, C. B., Gregory, R. T., and Felton, E. A., 1988, Evidence for low temperatures and biological diversity in Cretaceous high latitudes of Australia: Science, v. 242, p. 1403–1406.

Ripperdan, R. L., 1994, Global variations in carbon isotope composition during the latest Neoproterozoic and earliest Cambrian: Annual Review of Earth and Planetary Sciences, v. 22, p. 385–417.

Robardet, M., and Doré, F., 1988, The Late Ordovician diamictite formations from southwestern Europe: North Gondwana glaciomarine deposits: Palaeogeography, Palaeoclimatology, Palaeoecology, v. 66, p. 19–31.

Robb, L. J., Davis, D. W., and Kamo, S. L., 1991, Chronological framework for the Witwatersrand basin and environs: towards a time-constrained depositional model: South African Journal of Geology, v. 94, p. 86–95.

Roberts, J. D., 1976, Late Precambrian dolomites, Vendian glaciation, and synchroneity of Vendian glaciations: Journal of Geology, v. 84, p. 47–63.

Roberts, J., Claoué-Long, J. C., and Foster, C. B., 1996, SHRIMP zircon dating of the Permian System of eastern Australia: Australian Journal of Earth Sciences, v. 43, p. 401–421.

Robertson, A. H. F., Searle, M. P., and Ries, A. C., eds., 1990, The geology and tectonics of the Oman region: Geological Society of London, Special Publication 44, 845 p.

Robineau, B., and Ritz, M., 1990, Geoelectrical signature of the central Mauritanides deep structure, Mauritania, West Africa: Tectonics, v. 9, p. 1649–1661.

Robinson, J. M., 1990, Lignin, land plants, and fungi: Biological evolution affecting Phanerozoic oxygen balance: Geology, v. 15, p. 607–610.

R.O.C.C. Group (Research on Cretaceous Cycles Group), 1986, Rhythmic bedding in Upper Cretaceous pelagic carbonate sequences: Varying sedimentary response to climatic forcing: Geology, v. 14, p. 153–156.

Rocha-Campos, A. C., 1967, The Tubarão Group in the Brazilian portion of the Paraná basin, *in* Bigarella, J. J., Becker, R. D., and Pinto, I. D., eds., Problems in Brazilian Gondwana Geology: International symposium on the Gondwana stratigraphy and palaeontology: Curitiba, Brazil, p. 27–102.

Rocha-Campos, A. C., 1981, Late Ordovician(?)–Early Silurian, Trombetas Formation, Amazon Basin, Brazil, *in* Hambrey, M. J., and Harland, W. B., Earth's pre-Pleistocene glacial record: Cambridge, United Kingdom, Cambridge University Press, p. 896–898.

Rocha-Campos, A. C., and De Oliveira, M. E. C. B., 1972, Lower Gondwana rocks in Angola and Mozambique: Instituto de Investigação Científica de Angola Bulletin, v. 9, p. 51–74.

Rocha-Campos, A. C., and Hasui, Y., 1981, Late Precambrian Jangada Group and Puga Formation of central western Brazil; Proterozoic diamictites of western Minas Gerais and eastern Goiás, central Brazil; The Late Precambrian Bebedouro Formation, Bahia, Brazil; Late Precambrian Salobro Formation of Brazil; The Precambrian Carandaí Formation of southeastern Minas Gerais, Brazil; Tillites of the Macaúbas Group (Proterozoic) in central Minas Gerais and southern Bahia, Brazil, *in* Hambrey, M. J., and Harland, W. B., Earth's pre-Pleistocene glacial record: Cambridge, United Kingdom, Cambridge University Press, p. 916–938.

Rogers, A. W., 1902, On a glacial conglomerate in the Table Mountain Series: Philosophical Society of South Africa Transactions, v. 2, p. 236–242.

Rogers, J. J. W., 1996, A history of continents in the past three billion years: Journal of Geology, v. 104, p. 91–107.

Rona, P. A., 1973, Relations between rates of sediment accumulation on continental shelves, sea-floor spreading, and eustasy inferred from the central North Atlantic: Geological Society of America Bulletin, v. 84, p. 2851–2872.

Roots, C. G., and Parrish, R. R., 1988, Age of the Mount Harper volcanic complex, southern Ogilvie Mountains, Yukon, radiogenic age and isotope studies, Report 2: Ottawa, Canada Geological Survey of Canada Paper 88-2, p. 29–35.

Roscoe, S. M., and Card, K. D., 1993, The reappearance of the Huronian in Wyoming: rifting and drifting of ancient continents: Canadian Journal of Earth Sciences, v. 30, p. 2475–2480.

Ross, C. A., and Ross, J. R. P., 1985, Late Paleozoic depositional sequences are synchronous and worldwide: Geology, v. 13, p. 194–197.

Ross, C. A., and Ross, J. R. P., 1988, Late Paleozoic transgressive-regressive deposition, *in* Wilgus, C. K., Hastings, B. S., Kendall, C. G. St. C., Posamentier, H. W., Ross, C. A., and Van Wagoner, J. C., eds., Sea-level change: An integrated approach: Tulsa, Oklahoma, Society of Economic Paleontologists and Mineralogists Special Publication 42, p. 227–248.

Ross, C. A., Baud, A., and Menning, M., 1994, Pangea time scale, *in* Klein, G. D., ed., Pangea: Paleoclimate, tectonics, and sedimentation during accretion, zenith, and breakup of a supercontinent: Geological Society of America Special Paper 288, p. 10.

Ross, G. M., 1991, Tectonic setting of the Windermere Supergroup revisited: Geology, v. 19, p. 1125–1128.

Ross, G. M., Milkereit, B., Eaton, D., White, D., Kanasewich, E. R., and Burianyk, M. J. A., 1995, Paleoproterozoic collisional orogen beneath the western Canada sedimentary basin imaged by Lithoprobe crustal seismic-reflection data: Geology, v. 23, p. 195–199.

Ross, J. F., 1996, A few miles of land arose from the sea—and the world changed: Smithsonian, v. 27, no. 9, p. 112–121.

Ross, J. R. P., and Ross, C. A., 1992, Ordovician sea-level fluctuations, *in* Webby, B. D., and Laurie, J. R., eds., Global perspectives on Ordovician geology: Rotterdam, Netherlands, Balkema, p. 327–336.

Rubincam, D. P., 1995, Has climate changed the Earth's tilt?: Paleoceanography, v. 10, p. 365–372.

Ruddiman, W. F., ed., 1997, Tectonic uplift and climate change: New York, Plenum Press, 535 p.

Ruddiman, W. F., and Raymo, M. E., 1988, Northern Hemisphere climate régimes during the past 3 Ma: possible tectonic connections: Royal Society of London Philosophical Transactions, v. B318, p. 411–430.

Ruddiman, W. G., and Wright, H. E., Jr., eds., 1987, North America and adjacent oceans during the last deglaciation: Boulder, Colorado, Geological Society of America, The Geology of North America, v. K-3, 501 p.

Rui, Z. Q., and Piper, J. D. A., 1997, Palaeomagnetic study of Neoproterozoic glacial rocks of the Yangzi block: palaeolatitude and configuration of South China in the late Proterozoic supercontinent: Precambrian Research, v. 85, p. 173–199.

Runnegar, B., 1991, Precambrian oxygen levels estimated from the biochemistry and physiology of early eukaryotes: Global and planetary change, v. 5, p. 97–111.

Ruppel, S. C., James, E. W., Barrick, J. E., Nowlan, G., and Uyeno, T. T., 1996, High-resolution $^{87}Sr/^{86}Sr$ chemostratigraphy of the Silurian: Implications for event correlation and strontium flux: Geology, v. 24, p. 831–834.

Rust, I. C., 1975, Tectonic and sedimentary framework of Gondwana basins in southern Africa, *in* Campbell, K. S. W., ed., Gondwana Geology: Canberra, Australian National University Press, p. 537–564.

Rust, I. C., 1981, Early Palaeozoic Pakhuis Tillite, South Africa, *in* Hambrey, M. J., and Harland, W. B., eds., Earth's pre-Pleistocene glacial record: Cambridge, United Kingdom, Cambridge University Press, p. 113–117.

Rust, I. C., and Theron, J. N., 1964, Some aspects of the Table Mountain Series near Vanrhynsdorp: Geological Society of South Africa Transactions, v. 67, p. 133–137.

Sadowski, G. R., and Bettencourt, J. S., 1996, Mesoproterozoic tectonic correlations between eastern Laurentia and the western border of the Amazon craton: Precambrian Research, v. 76, p. 213–227.

Sæther, T., and Nystuen, J. P., 1981, Tectonic framework, stratigraphy, sedimentation and volcanism of the Late Precambrian Hedmark Group, Osterdalen, south Norway: Norges Geologiske Undersøkelse, v. 61, p. 193–211.

Sagan, C., and Chyba, C., 1997, The early faint sun paradox: Organic shielding of ultraviolet-labile greenhouse gases: Science, v. 276, p. 1217–1221.

Salop, L. J., 1977, Glaciation, rapid changes in organic evolution and their relationships with cosmic phenomena: International Geology Review, v. 19, p. 1271–1291.

Salop, L. J., 1983, Geological evolution of the Earth during the Precambrian: Berlin, Springer-Verlag, 459 p.

Sanders, J. E., 1995, Astronomical forcing functions: From Hutton to Milankovitch and beyond: Northeastern Geology and Environmental Sciences, v. 17, p. 306–345.

Savin, S. M., 1982, Stable isotopes in climatic reconstructions, *in* Climate in Earth History, Studies in Geophysics: Washington, D.C., National Academy Press, p. 164–171.

Saylor, B. Z., Grotzinger, J. P., and Germs, G. J. B., 1995, Sequence stratigraphy and sedimentology of the Neoproterozoic Kuibis and Schwarzrand Subgroups (Nama Group), southwestern Namibia: Precambrian Research, v. 73, p. 153–171.

Saylor, B. Z., Kaufman, A. J., Grotzinger, J. P., and Urban, F., 1998, A composite reference section for terminal Proterozoic strata of southern Namibia: Journal of Sedimentary Research, v. 68, p. 1223–1235.

Schenk, P. E., 1965, Depositional environment of the Gowganda Formation (Precambrian) at the south end of Lake Timagami, Ontario: Journal of Sedimentary Petrology, v. 35, p. 309–318.

Schenk, P. E., 1972, Possible Late Ordovician glaciation of Nova Scotia: Canadian Journal of Earth Sciences, v. 9, p. 95–107.

Schermerhorn, L. J. G., 1966, Terminology of mixed coarse-fine sediments: Journal of Sedimentary Petrology, v. 36, p. 831–835.

Schermerhorn, L. J. G., 1974, Late Precambrian mixtites: Glacial and/or non-glacial?: American Journal of Science, v. 274, p. 673–824.

Schermerhorn, L. J. G., 1976, Late Precambrian mixtites: Glacial and/or non-glacial?: Reply: American Journal of Science, v. 276, p. 375–384.

Schermerhorn, L. J. G., 1983, Late Proterozoic glaciation in the light of CO_2 depletion in the atmosphere, *in* Medaris, L. G., Jr., Byers, C. W., Mickelson, D. M., and Shanks, W. C., eds., Proterozoic geology: Selected papers from an international Proterozoic symposium: Geological Society of America Memoir 161, p. 309–315.

Schermerhorn, L. J. G., and Stanton, W. I, 1963, Tilloids in the West Congo Geosyncline: Geological Society of London Quarterly Journal, v. 119, p. 201–214.

Schidlowski, M., 1988, A 3,800 million-year isotopic record of life from carbon in sedimentary rocks: Nature, v. 333, p. 313–318.

Schidlowski, M., Hayes, J. M., and Kaplan, I. R., 1983, Isotopic inferences of ancient biochemistries: carbon, sulfur, hydrogen, and nitrogen, *in* Schopf, J. W., ed., Earth's earliest biosphere: Its origin and evolution: Princeton, New Jersey, Princeton University Press, p. 149–186.

Schlagintweit, O., 1943, La posición estratigráfica del yacimiento de hierro de Zapla y la difusión del horizonte glacial de Zapla en la Argentina y en Bolivia: Revista Minera, Geologia y Mineralogia, v. 13, p. 115–127.

Schmidt, P. W., and Williams, G. E., 1991, Palaeomagnetic correlation of the Acraman impact structure and the Late Proterozoic Bunyeroo eject horizon, South Australia: Australian Journal of Earth Sciences, v. 38, p. 283–289.

Schmidt, P. W., and Williams, G. E., 1995, The Neoproterozoic climatic paradox: Equatorial palaeolatitude for Marinoan glaciation near sea level in South Australia: Earth and Planetary Science Letters, v. 134, p. 107–124.

Schmidt, P. W., Williams, G. E., and Embleton, B. J. J., 1991, Low palaeolatitude of Late Proterozoic glaciation: early timing of remanence in haematite of the Elatina Formation, South Australia: Earth and Planetary Science Letters, v. 105, p. 355–367.

Schopf, J. W., ed., 1982, Earth's earliest biosphere: Its origin and evolution: Princeton, New Jersey, Princeton University Press, 543 p.

Schopf, J. W., 1992, Times of origin and earliest evidence of major biologic groups, *in* Schopf, J. W., and Klein, C., eds., The Proterozoic biosphere: New York, Cambridge University Press, p. 587–593.

Schopf, J. W., 1993, Microfossils of the Early Archean Apex Chert: New evidence of the antiquity of life: Science, v. 260, p. 640–646.

Schopf, J. W., 1994, Disparate rates, differing fates: Tempo and mode of evolution changed from the Precambrian to the Phanerozoic: National Academy of Sciences Proceedings, v. 91, p. 6735–6742.

Schopf, J. W., and Klein, C., eds., 1992, The Proterozoic biosphere: New York, Cambridge University Press, 1348 p.

Schopf, T. J. M., 1980, Paleoceanography: Cambridge, Massachusetts, Harvard University Press, 341 p.

Schwarzbach, M., 1976, Late Paleozoic glaciation: Part VI, Asia: Discussion, Geological Society of America Bulletin, v. 87, p. 640.

Sclater, J. G., and Francheteau, J., 1970, The implications of terrestrial heat flow observations on current tectonic and geochemical models of the crust and upper mantle of the earth: Royal Astronomical Society Geophysical Journal, v. 20, p. 509–542.

Sclater, J. G., Hellinger, S., and Tapscott, C., 1977, The paleobathymetry of the Atlantic Ocean from the Jurassic to the present: Journal of Geology, v. 85, p. 509–552.

Scoates, J. S., Frost, C. D., Mitchell, J. N., Lindsley, D. H., and Frost, B. R., 1996, Residual-liquid origin for a monzonitic intrusion in a mid-Proterozoic anorthosite complex: The Sybille intrusion, Laramie anorthosite complex, Wyoming: Geological Society of America Bulletin, v. 108, p. 1357–1371.

Scotese, C. R., 1990, Atlas of Phanerozoic plate tectonic reconstructions, International Lithosphere Program (IUGG-IUGS), Paleomap Project: Arlington, Texas, Department of Geology, University of Texas, p. 1–35, 53 paleomaps.

Seinfeld, J. H., 1998, Clouds, contrails and climate: Nature, v. 391, p. 837–838.

Selwyn, A. R. C., 1859, Geological notes of a journey in South Australia from Cape Jervis to Mount Serle: Adelaide Parliamentary Papers, South Australia, v. 20, 4 p.

Şengör, A. M. C., 1987, Tectonics of the Tethysides: Orogenic collage development in a collisional setting: Annual Review of Earth and Planetary Sciences, v. 15, p. 213–244.

Şengör, A. M. C., Altiner, D., Cin, A., Ustaömer, T., and Hsü, K. J., 1988, Origin and assembly of the Tethyside orogenic collage at the expense of Gondwana Land, *in* Audley-Charles, M. G., and Hallam, A., eds., Gondwana and Tethys: Oxford, United Kingdom, Oxford University Press, Geological Society of London Special Publication 37, p. 119–181.

Shackleton, N. J., 1987a, Oxygen isotopes, ice volume, and sea level: Quaternary Science Reviews, v. 6, p. 183–190.

Shackleton, N. J., 1987b, The carbon isotope record of the Cenozoic history of organic carbon burial and of oxygen in the ocean and atmosphere, *in* Brooks, J., and Fleet, A. J., eds., Marine Petroleum Source Rocks: Geological Society of London Special Publication 26, p. 423–434.

Shackleton, N. J., and Opdyke, N. D., 1973, Oxygen isotope and paleomagnetic stratigraphy of equatorial Pacific core V28-238: oxygen isotope temperatures and ice volume on a 100,000 and 1,000,000 year scale: Quaternary Research, v. 3, p. 39–55.

Shackleton, R. M., Ries, A. C., Coward, M. P., and Cobbold, P. R., 1979, Structure, metamorphism and geochronology of the Arequipa massif of coastal Peru: Geological Society of London Journal, v. 136, p. 195–214.

Shaw, H. R., 1987, The periodic structure of the natural record, and nonlinear dynamics: EOS (American Geophysical Union Transactions), v. 68, p. 1651–1665.

Shear, W. A., 1991, The early development of terrestrial ecosystems: Nature, v. 351, p. 283–289.

Sheehan, P. M. 1973, The relation of Late Ordovician glaciation to the Ordovician-Silurian changeover in North American brachiopod fauna: Lethaia, v. 6, p. 147–154.

Sheldon, R. P., 1981, Ancient marine phosphorites: Annual Review of Earth and Planetary Sciences, v. 9, p. 251–284.

Sheldon, R. P., 1984, Ice-ring origin of the Earth's atmosphere and hydrosphere and Late Proterozoic–Cambrian phosphogenesis: Geological Survey of India Special Publication 17, p. 17–21.

Shih-fan Liao, 1981, Sinian glacial deposits of Guizhou Province, China, *in* Hambrey, M. J., and Harland, W. B., eds., Earth's pre-Pleistocene glacial record: Cambridge, United Kingdom, Cambridge University Press, p. 414–423.

Shih-fan Liao, Wan yen-gen, and research group of Sinian glaciation, 1976, Sinian glacial deposits of China: Peking, China, p. 1–15.

Shoemaker, E. M., and Wolfe, R. F., 1986, Mass extinctions, crater ages and comet showers, *in* Smoluchowski, R., Bahcall, J. N., and Matthews, M. S., eds., The Galaxy and the Solar System: Tucson, Arizona, University of Arizona Press, p. 338–386.

Shoemaker, E. M., Wolfe, R. F., and Shoemaker, C. S., 1990, Asteroid and comet flux in the neighborhood of Earth, *in* Sharpton, V. L., and Ward, P. D., eds., Global catastrophes in Earth history: An interdisciplinary conference on impacts, volcanism, and mass mortality: Geological Society of America Special Paper 247, p. 155–170.

Siedlecka, A., and Roberts, D., 1992, The bedrock geology of the Varanger Peninsula, Finnmark, North Norway: an excursion guide: Oslo, Norway, Norges Geologiske Undersøkelse Special Publication 5, p. 1–45.

Silver, L. T., and Schultz, P. H., eds., 1982, Geological implications of impacts of large asteroids and comets on the Earth: Geological Society of America Special Paper 190, 528 p.

Singh, T., 1987, Permian biogeography of the Indian subcontinent with special reference to the marine fauna, *in* McKenzie, G. D., ed., Gondwana six: Stratigraphy, sedimentology, and paleontology: Washington, D.C., American Geophysical Union, Geophysical Monograph 41 p. 239–249.

Slingo, A., 1990, Sensitivity of the Earth's radiation budget to changes in low clouds: Nature, v. 343, p. 49–51.

Socci, A. D., 1992, Climate, glaciation and deglaciation: controls, pathways, feedbacks, rates and frequencies: Modern Geology, v. 16, p. 279–316.

Sohl, L. E., 1997, Paleomagnetic and stratigraphic implications for the duration of low-latitude glaciation in the late Neoproterozic of Australia: Geological Society of America Abstracts with Programs, v. 29, no. 6, p. A–195.

Sohl, L. E., and Christie-Blick, N., 1995, Equatorial glaciation in the Neoproterozoic: New evidence from the Marinoan glacial succession of Australia: Geological Society of America Abstracts with Programs, v. 27, no. 6, p. A204.

Söhnge, A. P. G., 1984, Glacial diamictite in the Peninsula Formation near Cape Hangklip: Geological Society of South Africa Transactions, v. 87, p. 199–210.

Sonett, C. P., Finney, S. A., and Williams, C. R., 1988, The lunar orbit in the late Precambrian and the Elatina sandstone laminae: Nature, v. 335, p. 806–808.

Sonett, C. P., Kvale, E. P., Zakharian, A., Chan, M. A., and Demko, T. M., 1996, Late Proterozoic and Paleozoic tides, retreat of the moon, and rotation of the Earth: Science, v. 273, p. 100–104.

Soreghan, G. S., and Giles, K. A., 1999, Amplitudes of Late pennsylvanian glacioeustasy: Geology, v. 27, p. 255–258.

Spencer, A. M., 1971, Late Pre-Cambrian glaciation in Scotland: Geological Society of London Memoir 6, 98 p.

Spirakis, C. S., 1989, Possible effect of readily available iron in volcanic ash on the carbon to sulfur ratio in lower Paleozoic normal marine sediments and implications for atmospheric oxygen: Geology, v. 17, p. 599–601.

Spjeldnæs, N., 1961, Ordovician climatic zones: Norsk Geologisk Tidskrift, v. 41, p. 45–77.

Spjeldnæs, N., 1964, The Eocambrian glaciation in Norway: Geologische Rundschau, v. 54, p. 24–45.

Spjeldnæs, N., 1981, Lower Palaeozoic palaeoclimatology, in Holland, C. H., ed., Lower Palaeozoic of the Middle East, eastern and southern Africa, and Antarctica: New York, John Wiley and Sons, p. 199–256.

Stanistreet, I. G., and McCarthy, T. S., 1991, Changing tectono-sedimentary scenarios relevant to the development of the Late Archaean Witwatersrand basin: Journal of African Earth Sciences, v. 13, p. 65–81.

Stanistreet, I. G., Kukla, P. A., Henry, G., 1991, Sedimentary basinal responses to a Late Precambrian Wilson cycle: the Damara orogen and Nama foreland, Namibia: Journal of African Earth Sciences, v. 13, p. 141–156.

Stanley, S. M., 1988, Paleozoic mass extinctions: shared patterns suggest global cooling as a common cause: American Journal of Science, v. 288, p. 334–352.

Stanley, S. M., 1995, New horizons for paleontology, with two examples: The rise and fall of the Cretaceous Supertethys and the cause of the modern ice age: Journal of Paleontology, v. 69, p. 999–1007.

Stanley, S. M., and Yang, X., 1994, A double mass extinction at the end of the Paleozoic Era: Science, v. 266, p. 1340–1344.

Stauffer, P. H., and Lee, C. P., 1987, The Upper Paleozoic pebbly mudstone facies of peninsular Thailand and western Malaysia—continental margin deposits of Palaeoeurasia—discussion of Altermann, W., 1986: Geologische Rundschau, v. 76, p. 945–948.

Steiger, R. H., and Jäger, E., 1977, Subcommission on geochronology: Convention of the use of decay constants in geo- and cosmochronology: Earth and Planetary Science Letters, v. 36, p. 359–362.

Steiner, J., and Falk, F., 1981, The Ordovician Lederschiefer of Thuringia, in Hambrey, M. J., and Harland, W. B., eds., Earth's pre-Pleistocene glacial record: Cambridge, United Kingdom, Cambridge University Press, p. 579–581.

Steiner, J., and Grillmair, E., 1973, Possible galactic causes of periodic and episodic glaciations: Geological Society of America Bulletin, v. 84, p. 1003–1018.

Stern, R. J., 1994, Arc assembly and continental collision in the Neoproterozoic east African orogen: Implications for the consolidation of Gondwanaland: Annual Review of Earth and Planetary Sciences, v. 22, p. 319–351.

Stern, T. A., and Ten Brink, U.S., 1989, Flexural uplift of the Transantarctic Mountains: Journal of Geophysical Research, v. 94, p. 10315–10330.

Stewart, A. D., 1997, Discussion on indications of glaciation at the base of the Proterozoic Stoer Group (Torridonian), NW Scotland: Geological Society of London Journal, v. 154, p. 375–376.

Stoll, H. M., and Schrag, D. P., 1996, Evidence for glacial control of rapid sea level changes in the early Cretaceous: Science, v. 272, p. 1771–1774.

Stratten, T., 1969, A preliminary report of a directional study of the Dwyka Tillites in the Karroo Basin of South Africa, in Gondwana Stratigraphy, v. 2: Paris, United Nations Educational, Scientific and Cultural Organization, p. 741–762.

Stump, E., and Fitzgerald, P. G., 1992, Episodic uplift of the Transantarctic Mountains: Geology, v. 20, p. 161–164.

Stump, E., Miller, J. M. G., Korsch, R. J., and Edgerton, D. G., 1988, Diamictite from Nimrod Glacier area, Antarctica: Possible Proterozoic glaciation on the seventh continent: Geology, v. 16, p. 225–228.

Suárez-Soruco, R., 1995, Comentarios sobre la edad de la Formacion Cancañari: Revista Tecnica de Yacimientos Petroliferos Fiscalels Bolivianos, v. 16, p. 51–54.

Sugden, D. E., and John, B. S., 1976, Glaciers and landscape: A geomorphological approach: London, Edward Arnold, 376 p.

Sun Dong-li, 1993, On the Permian biogeographic boundary between Gondwana and Eurasian Tibet, China, as the eastern section of the Tethys: Palaeogeography, Palaeoclimatology, Palaeoecology, v. 100, p. 59–77.

Sundquist, E. T., 1993, The global carbon dioxide budget: Science, v. 259, p. 934–941.

Sundquist, E. T., and Broecker, W. S., eds., 1985, The carbon cycle and atmospheric CO_2: Natural variations Archean to present: Washington, D.C., American Geophysical Union, Geophysical Monograph 32, 625 p.

Surlyk, F., 1991, Tectonostratigraphy of North Greenland: Grønlands Geologiske Undersøgelse Bulletin 160, p. 25–47.

Sutherland, P. C., 1870, Notes on an ancient boulder-clay of Natal: Geological Society of London Quarterly Journal, v. 26, p. 514.

Sutherland, P. C., 1871, Notes on an ancient boulder-clay of Natal: Geological Society of London Quarterly Journal, v. 27, p. 39.

Sutton, J., and Watson, J., 1954, Ice-borne boulders in the Macduff Group of the Dalradian of Banffshire: Geological Magazine, v. 91, p. 391–398.

Symonds, R. B., Rose, W. I., and Reed, M. H., 1988, Contribution of Cl- and F-bearing gases to the atmosphere by volcanoes: Nature, v. 334, p. 415–418.

Talbot, M. R., 1981, Early Palaeozoic(?) diamictites of south-west Ghana, in Hambrey, M. J., and Harland, W. B., eds., Earth's pre-Pleistocene glacial record: Cambridge, United Kingdom, Cambridge University Press, p. 108–112.

Tankard, A. J., Jackson, M. P. A., Eriksson, K. A., Hobday, D. K., Hunter, D. R., and Minter, W. E. L., 1982, Crustal evolution of southern Africa: 3.8 billion years of Earth history: New York, Springer-Verlag, 523 p.

Tans, P. P., Fung, I. Y., and Takahashi, T., 1990, Observational constraints on the global atmospheric CO_2 budget: Science, v. 247, p. 1431–1438.

Tarling, D. H., 1974, A palaeomagnetic study of Eocambrian tillites in Scotland: Geological Society of London Quarterly Journal, v. 130, p. 163–177.

Taylor, A. D., Baggaley, W. J., and Steel, D. I., 1996, Discovery of interstellar dust entering the Earth's atmosphere: Nature, v. 380, p. 323–325.

Taylor, K. E., and Penner, J. E., 1994, Response of the climate system to atmospheric aerosols and greenhouse gases: Nature, v. 369, p. 734–737.

Taylor, S. R., 1992, Vestiges of a beginning: Nature, v. 360, p. 710–711.

Taylor, S. R., 1998, Destiny or chance, our solar system and its place in the cosmos: Cambridge, United Kingdom, Cambridge University Press, 229 p.

Thelander, T., 1981, The late Precambrian Längmarkberg Formation in the central Swedish Caledonides, in Hambrey, M. J., and Harland, W. B., eds., Earth's pre-Pleistocene glacial record: Cambridge, United Kingdom, Cambridge University Press, p. 615–619.

Theron, J. N., 1994, The Ordovician system in South Africa: International Union of Geological Sciences Publication No. 29, B, p. 1–5.

Thistlewood, L., Leat, P. T., Millar, I. L., Storey, B. C., and Vaughan, A. P. M., 1997, Basement geology and Palaeozoic-Mesozoic mafic dykes from the Cape Meredith complex, Falkland Islands: a record of repeated intracontinental extension: Geological Magazine, v. 134, p. 355–367.

Thomas, G. S. P., and Connell, R. J., 1985, Iceberg drop, dump, and grounding structures from Pleistocene glacio-lacustrine sediments, Scotland: Journal of Sedimentary Petrology, v. 55, p. 243–249.

Thomson, J., 1871, On the occurrence of pebbles and boulders of granite in schistose rocks in Islay, Scotland, *in* Transactions, 40th meeting of the British Association: Liverpool, p. 88.

Thomson, J., 1877, On the geology of the Island of Islay: Geological Society of Glasgow Transactions, v. 5, p. 200–222.

Thomson, W. (Lord Kelvin), 1862, On the secular cooling of the earth: Royal Society of Edinburgh Transactions, v. 23, p. 157–169. (Reprinted in Thomson and Tait, 1890, Treatise on natural philosophy, Part II: Cambridge, United Kingdom, Cambridge University Press, p. 468–485.)

Torsvik, T. H., Lohmann, K. C., and Sturt, B. A., 1995, Vendian glaciations and their relation to the dispersal of Rodinia: Paleomagnetic constraints: Geology, v. 23, p. 727–730.

Torsvik, T. H., Smethurst, M. A., Meert, J. G., Van der Voo, R., McKerrow, W. S., Brasier, M. D., Sturt, B. A., and Walderhaug, H. J., 1996, Continental break-up and collision in the Neoproterozoic and Palaeozoic—A tale of Baltica and Laurentia: Earth Science Reviews, v. 40, p. 229–259.

Totten, S. M., and White, G. W., 1985, Glacial geology and the North American craton: Significant concepts and contributions of the nineteenth century, *in* Drake, E. T., and Jordan, W. M., eds., Geologists and ideas: A history of North American geology: Boulder, Colorado, Geological Society of America, Centennial Special Volume 1, p. 125–141.

Treagus, J. E., 1981, The Lower Kinlochlaggan Boulder Bed, central Scotland, *in* Hambrey, M. J., and Harland, W. B., eds., Earth's pre-Pleistocene glacial record: Cambridge, United Kingdom, Cambridge University Press, p. 637–639.

Trendall, A. F., 1976, Striated and faceted boulders from the Turee Creek Formation—evidence for a possible Huronian glaciation on the Australian continent: Perth, Western Australia, Geological Survey of Western Australia Annual Report for the year 1975, p. 89–92.

Trendall, A. F., 1981, The lower Proterozoic Meteorite Bore Member, Hamersley Basin, Western Australia, *in* Hambrey, M. J., and Harland, W. B., eds., Earth's pre-Pleistocene glacial record: Cambridge, United Kingdom, Cambridge University Press, p. 555–557.

Trettin, H. P., 1989, The Arctic Islands, *in* Bally, A. W., and Palmer, A. R., eds., The Geology of North America—An overview: Boulder, Colorado, Geological Society of America, The Geology of North America, v. A, p. 349–370.

Trompette, R., 1973, Le Précambrien supérieur et le Paléozoïque inférieur de l'Adrar de Mauritanie (bordure occidentale du basin de Taoudeni, Afrique de l'Ouest): Un exemple de sédimentation de craton: Travaux des Laboratoires des Sciences de la Terre, Saint-Jérôme, Marseille, B, v. 7, 702 p.

Trompette, R., 1981, Late Precambrian tillites of the Volta Basin and the Dahomeyides orogenic belt (Benin, Ghana, Niger, Togo and Upper Volta), *in* Hambrey, M. J., and Harland, W. B., eds., Earth's pre-Pleistocene glacial record: Cambridge, United Kingdom, Cambridge University Press, p. 135–139.

Trompette, R., 1982, Upper Proterozoic (1800–570 Ma) stratigraphy: a survey of lithostratigraphic, paleontological, radiochronological, and magnetic correlations: Precambrian Research, v. 18, p. 27–52.

Trompette, R., 1994, Geology of western Gondwana (2000–500 Ma): Pan-African–Brasiliano aggregation of South America and Africa: Rotterdam, Netherlands, Balkema, 350 p.

Trompette, R., 1996, Temporal relationship between cratonization and glaciation: The Vendian–early Cambrian glaciation in Western Gondwana: Palaeogeography, Palaeoclimatology, Palaeoecology, v. 123, p. 373–383.

Truswell, E. M., 1980, Permo-Carboniferous palynology of Gondwanaland: progress and problems in the decade to 1980: Australia Bureau of Mineral Resources Journal of Geology and Geophysics, v. 5, p. 95–111.

Tucker, M. E., and Reid, P. C., 1973, The sedimentology and context of Late Ordovician glacial marine sediments from Sierra Leone, West Africa: Palaeogeography, Palaeoclimatology, Palaeoecology, v. 13, p. 289–307.

Tucker, M. E., and Reid, P. C., 1981, Late Ordovician glaciomarine sediments, Sierra Leone, *and* Late Precambrian glacial sediments, Sierra Leone, *in* Hambrey, M. J., and Harland, W. B., eds., Earth's pre-Pleistocene glacial record: Cambridge, United Kingdom, Cambridge University Press, p. 97–98, 132–134.

Umhoefer, P. J., Dorsey, R. J., and Renne, P., 1994, Tectonics of the Pliocene Loreto basin, Baja California Sur, Mexico, and evolution of the Gulf of California: Geology, v. 22, p. 649–652.

Underwood, C. J., Deynoux, M., and Ghienne, J.-F., 1998, High palaeolatitude (Hodh, Mauritania) recovery of graptolite faunas after the Hirnantian (end Ordovician) extinction event: Palaeogeography, Palaeoclimatology, Palaeoecology, v. 142, p. 91–105.

Unrug, R., 1991, The Mwembeshi and Zambesi dislocation systems: The central segment of a transcontinental shear zone in south-central Africa, *in* Ulrich, H., and Rocha-Campos, A. C., eds., Gondwana Seven, Proceedings: São Paulo, Brasil, Instituto de Geociências, p. 57–64.

Unrug, R., 1996, The assembly of Gondwana: Episodes, v. 19, p. 11–20.

Unrug, R., 1997, Rodinia to Gondwana: The geodynamic map of Gondwana supercontinent assembly: GSA Today, v. 7, p. 1–6.

Urey, H. C., 1947, The thermodynamic properties of isotopic substances: Journal of Chemical Society (London), 1947, p. 562–581.

USDMA Chart, 1990, Pilot chart of the North Atlantic Ocean: U.S. Defense Mapping Agency, DMA Stock Number, Pilot 169010 (October 1990).

Vail, P. R., Mitchum, R. M., Jr., Todd, R. G., Widmier, J. M., Thompson, S., III, Sangree, J. B., Bubb, J. N., and Hatlelid, W. G., 1977, Seismic stratigraphy and global changes of sea level, *in* Payton, C. E., ed., Seismic stratigraphy—Application to hydrocarbon explanation: Tulsa, Oklahoma, American Association of Petroleum Geologists Memoir 26, p. 49–212.

Van Cappellen, P., and Ingall, E. D., 1996, Redox stabilization of the atmosphere and oceans by phosphorus-limited marine productivity: Science, v. 271, p. 493–496.

Van der Gracht, W. A. J. M. V. W., ed., 1928, Theory of continental drift: Tulsa, Oklahoma, American Association of Petroleum Geologists, 240 p.

Van der Voo, R., 1993, Paleomagnetism of the Atlantic, Tethys, and Iapetus oceans: London, Cambridge University Press, 411 p.

Van der Voo, R., and Meert, J. G., 1991, Late Proterozoic paleomagnetism and tectonic models: a critical appraisal: Precambrian Research, v. 53, p. 149–163.

Van Schmus, R., 1965, The geochronology of the Blind River–Bruce mines area, Ontario, Canada: Journal of Geology, v. 73, p. 755–780.

Van Schmus, R., 1980, Geochronology of igneous rocks associated with the Penokean orogeny in Wisconsin, *in* Morey, G. B., and Hanson, G. N., eds., Selected studies of Archean gneisses and lower Proterozoic rocks, southern Canadian Shield: Geological Society of America Special Paper 182, p. 159–168.

Van Schmus, W. R., Bickford, M. E., Anderson, J. L., Bender, E. E., Anderson, R. R., Bauer, P. W., Robertson, J. M., Bowring, S. A., Condie, K. C., Denison, R. E., Gilbert, M. C., Grambling, J. A., Mawer, C. K., Shearer, C. K., Hinze, W. J., Karlstrom, K. E., Kisvarsanyi, E. B., Lidiak, E. G., Reed, J. C., Jr., Sims, P. K., Tweto, O., Silver, L. T., Treves, S. B., Williams, M. L., Wooden, J. L., 1993, Transcontinental Proterozoic provinces, *in* Reed, J. C., Jr., Bickford, M. E., Houston, R. S., Link, P. K., Rankin, D. W., Sims, P. K., and Van Schmus, W. R., eds., Precambrian: Conterminous U.S.: Boulder, Colorado, Geological Society of America, The Geology of North America, v. C-2, p. 171–334.

Van Schmus, W. R., and Hinze, W. J., 1985, The midcontinent rift system: Annual Review of Earth and Planetary Sciences, v. 13, p. 345–383.

Van Veen, P. M., and Simonsen, B. T., 1991, Glacial-eustatic sea-level curve for early Late Pennsylvanian sequence in north-central Texas and biostratigraphic correlation with curve for midcontinent North America: Comment: Geology, v. 20, p. 91–94.

Vanyo, J. P., and Awramik, S. M., 1982, Length of day and obliquity of the ecliptic 850 Ma ago: Preliminary results of a stromatolite growth model: Geophysical Research Letters, v. 9, p. 1125–1128.

Vaslet, D., 1990, Upper Ordovician glacial deposits in Saudi Arabia: Episodes, v. 13, p. 147–161.

Vavrdová, M., Isaacson, P. E., Díaz-Martinez, E. D., and Bek, J., 1991, Palinologia del limite Devonico-Carbonifero entorno al Lago Titikaka, Bolivia: Resultados preliminares, Revista Tecnica de Yacimientos Petroliferos Fiscales Bolivianos, v. 12, p. 303–315.

Veevers, J. J., 1988, Gondwana facies started when Gondwanaland merged in

Pangea: Geology, v. 16, p. 732–734.

Veevers, J. J., 1990, Tectonic-climatic supercycle in the billion-year plate-tectonic eon: Permian Pangean icehouse alternates with Cretaceous dispersed-continents greenhouse: Sedimentary Geology, v. 68, p. 1–16.

Veevers, J. J., 1994a, Pangea: Evolution of a supercontinent and its consequences for Earth's paleoclimate and sedimentary environments, *in* Klein, G. D., ed., Pangea: Paleoclimate, tectonics, and sedimentation during accretion, zenith, and breakup of a supercontinent: Geological Society of America Special Paper 288, p. 13–23.

Veevers, J. J., 1994b, Case for the Gamburtsev subglacial mountains of East Antarctica originating by mid-Carboniferous shortening of an intracratonic basin: Geology, v. 22, p. 593–596.

Veevers, J. J., and Collinson, J. W., 1994, Late Carboniferous-Jurassic correlation chart for Antarctica, *in* Veevers, J. J., and Powell, C. McA., eds., Permian-Triassic Pangean basins and foldbelts along the Panthalassan margin of Gondwanaland: Geological Society of America Memoir 184, p. 209–213.

Veevers, J. J., and Powell, C. McA., 1987, Late Paleozoic glacial episodes in Gondwanaland reflected in transgressive-regressive depositional sequences in Euramerica: Geological Society of America Bulletin, v. 98, p. 475–487.

Veevers, J. J., and Powell, C. McA., eds., 1994, Permian-Triassic Pangean basins and foldbelts along the Panthalassan margin of Gondwanaland: Geological Society of America Memoir 184, 368 p.

Veevers, J. J., and Tewari, R. C., 1995, Gondwana master basin of Peninsular India between Tethys and the interior of the Gondwanaland province of Pangea: Geological Society of America Memoir 187, 72 p.

Veevers, J. J., Conaghan, P. J., and Powell, C. McA., 1994a, Eastern Australia, *in* Veevers, J. J., and Powell, C. McA., eds., Permian-Triassic Pangean basins and foldbelts along the Panthalassan margin of Gondwanaland: Geological Society of America Memoir 184, p. 11–171.

Veevers, J. J., Cole, D. I., and Cowan, E. J., 1994b, Southern Africa: Karoo basin and Cape fold belt, *in* Veevers, J. J., and Powell, C. McA., eds., Permian-Triassic Pangean basins and foldbelts along the Panthalassan margin of Gondwanaland: Geological Society of America Memoir 184, p. 223–279.

Veevers, J. J., Powell, C. McA., Collinson, J. W., and López-Gamundí, O. R., 1994c, Synthesis, *in* Veevers, J. J., and Powell, C. McA., eds., Permian-Triassic Pangean basins and foldbelts along the Panthalassan margin of Gondwanaland: Geological Society of America Memoir 184, p. 331–353.

Veevers, J. J., Conaghan, P. J., and Shaw, S. E., 1994d, Turning point in Pangean environmental history at the Permian/Triassic (P/Tr) boundary, *in* Klein, G. D., ed., Pangea: Paleoclimate, tectonics, and sedimentation during accretion, zenith, and breakup of a supercontinent: Geological Society of America Special Paper 288, p. 187–196.

Veizer, J., and Compston, W., 1974, $^{87}Sr/^{86}Sr$ composition of seawater during the Phanerozoic: Geochimica et Cosmochimica Acta, v. 38, p. 1461–1484.

Veizer, J., and Compston, W., 1976, $^{87}Sr/^{86}Sr$ in Precambrian carbonates as an index of crustal evolution: Geochimica et Cosmochimica Acta, v. 40, p. 905–914.

Verhoogen, J., Turner, F. J., Weiss, L. E., Wahrhafting, C., and Fyfe, W. S., 1970, The Earth: An introduction to physical geology: New York, Holt, Rinehart and Winston, Inc., 748 p.

Vermeij, G. J., 1993, The biological history of a seaway: Science, v. 260, p. 1603–1604.

Vidal, G., and Bylund, G., 1981, Micropalaeontology and palaeomagnetism of the tillite-bearing Proterozoic to Lower Palaeozoic sequences in Finnmark (Northery Norway), *in* Hambrey, M. J., and Harland, W. B., eds., Earth's pre-Pleistocene glacial record: Cambridge, United Kingdom, Cambridge University Press, p. 610.

Vidal, G., and Moczydłowska, M., 1995, The Neoproterozoic of Baltica—stratigraphy, palaeobiology and general geological evolution: Precambrian Research, v. 73, p. 197–216.

Villeneuve, M. 1989, The geology of the Madina-Kouta basin (Guinea-Senegal) and its significance for the geodynamic evolution of the western part of the West African craton during the Upper Proterozoic period: Precambrian Research, v. 44, p. 305–322.

Visser, J. N. J., 1971, The deposition of the Griquatown glacial member in the Transvaal Supergroup: Geological Society of South Africa Transactions, v. 74, p. 186–199.

Visser, J. N. J., 1981, The Mid-Precambrian tillite in the Griqualand West and Transvaal basins, South Africa, *in* Hambrey, M. J., and Harland, W. B., eds., Earth's pre-Pleistocene glacial record: Cambridge, United Kingdom, Cambridge University Press, p. 180–184.

Visser, J. N. J., 1982, Upper Carboniferous glacial sedimentation in the Karoo basin near Prieska, South Africa: Palaeogeography, Palaeoclimatology, Palaeoecology, v. 38, p. 63–92.

Visser, J. N. J., 1983a, Submarine debris flow deposits from the Upper Carboniferous Dwyka Tillite Formation in the Kalahari basin, South Africa: Sedimentology, v. 30, p. 511–523.

Visser, J. N. J., 1983b, Glacial marine sedimentation in the late Paleozoic Karoo basin, southern Africa, *in* Molnia, B. F., ed., Glacial-marine sedimentation: New York, Plenum Press, p. 667–701.

Visser, J. N. J., 1987, The palaeogeography of part of southwestern Gondwana during the Permo-Carboniferous glaciation: Palaeogeography, Palaeoclimatology, Palaeoecology, v. 61, p. 205–219.

Visser, J. N. J., 1990, The age of the late Palaeozoic glacigene deposits in southern Africa: South African Journal of Geology, v. 93, p. 366–375.

Visser, J. N. J., 1993a, Sea-level changes in a back-arc-foreland transition: the late Carboniferous-Permian Karoo basin of South Africa: Sedimentary Geology, v. 83, p. 115–131.

Visser, J. N. J., 1993b, Submarine debris flow deposits from the Upper Carboniferous Dwyka Tillite formation in the Kalahari basin, South Africa: Sedimentology, v. 30, p. 511–523.

Visser, J. N. J., 1994a, The interpretation of massive rain-out and debris-flow diamictites from the glacial marine environment, *in* Deynoux, M., Miller, J. M. G., Domack, E. W., Eyles, N., Fairchild, I. F., and Young, G. M., eds., Earth's glacial record: Cambridge, United Kingdom, Cambridge University Press, p. 83–94.

Visser, J. N. J., 1994b, A Permian argillaceous syn- to post-glacial foreland sequence in the Karoo basin, South Africa, *in* Deynoux, M., Miller, J. M. G., Domack, E. W., Eyles, N., Fairchild, I. F., and Young, G. M., eds., Earth's glacial record: Cambridge, United Kingdom, Cambridge University Press, p. 193–203.

Visser, J. N. J., 1997a, Deglaciation sequences in the Permo-Carboniferous Karoo and Kalahari basins of southern Africa: a tool in the analysis of cyclic glaciomarine basin fills: Sedimentology, v. 44, p. 507–521.

Visser, J. N. J., 1997b, A review of the Permo-Carboniferous glaciation in Africa, *in* Martini, I. P., ed., Late glacial and postglacial environmental changes: New York, Oxford University Press, p. 169–191.

Visser, J. N. J., and Hall, K. J., 1985, Boulder beds in the glaciogenic Permo-Carboniferous Dwyka Formation in South Africa: Sedimentology, v. 32, p. 281–294.

Visser, J. N. J., and Loock, J. C., 1982, An investigation of the basal Dwyka tillite in the southern part of the Karoo Basin, South Africa: Geological Society of South Africa Transactions, v. 85, p. 179–187.

Visser, J. N. J., van Niekerk, B. N., and van der Merwe, S. W., 1997, Sediment transport of the Late Paleozoic Dwyka Group in the southwestern Karoo Basin: South African Journal of Geology, v. 100, p. 223–236.

Volney, C. F., 1803, Tableau du climat et du sol des Etats Unis d'Amérique: Paris, Courcier, Dentu, An. XII-1803, 2 v., 300 p. and 532 p.

Volney, C. F., 1804, A view of the soil and climate of the United States of America: C. G. Brown, trans.: Philadelphia, J. Conrad and Co., 446 p. (Reprinted, with introduction by G. W. White, New York, Hafner Publishing Company, Inc., 1968.)

Von Brunn, V., 1987, A facies analysis of Permo-Carboniferous glacigene deposits along a paleoscarp in northern Natal, South Africa, *in* McKenzie, G., ed., Gondwana Six: Structure, Tectonics, and Geophysics: Washington, D.C., American Geophysical Union, Geophysical Monograph 41, p. 113–122.

Von Brunn, V., 1994, Glaciogenic deposits of the Permo-Carboniferous Dwyka Group in the eastern region of the Karoo Basin, South Africa, *in* Deynoux, M., Miller, J. M. G., Domack, E. W., Eyles, N., Fairchild, I. J., and Young,

G. M.: Cambridge, United Kingdom, Cambridge University Press, p. 60–69.

Von Brunn, V., and Gold, D. J. C., 1993, Diamictite in the Archaean Pongola sequence of southern Africa: Journal of African Earth Sciences, v. 16, p. 367–374.

Von Brunn, V., and Gravenor, C. P., 1983, A model for late Dwyka glaciomarine sedimentation in the eastern Karoo basin: Geological Society of South Africa Transactions, v. 86, p. 199–209.

Von Brunn, V., and Marshall, C. G. A., 1989, Glaciated surfaces and the base of the Dwyka Formation near Pietermaritzburg, Natal: South African Journal of Geology, v. 92, p. 420–426.

Von Brunn, V., and Stratten, T., 1981, Late Paleozoic tillites of the Karoo basin of South Africa, in Hambrey, M. J., and Harland, W. B., eds., Earth's pre-Pleistocene glacial record: Cambridge, United Kingdom, Cambridge University Press, p. 71–79.

Walker, J. C. G., 1977, Evolution of the atmosphere: New York, Macmillan Publishing Co., 318 p.

Walker, J. C. G., Klein, C., Schidlowski, M., Schopf, J. W., Stevenson, D. J., and Walter, M. R., 1983, Environmental evolution of the Archean–early Proterozoic Earth, in Schopf, J. W., ed., Earth's earliest biosphere: Its origin and evolution: Princeton, New Jersey, Princeton University Press, p. 260–290.

Walker, J. D., Klepacki, D. W., and Burchfiel, B. C., 1986, Late Precambrian tectonism in the Kingston Range, southern California: Geology, v. 14, p. 15–18.

Wallace, P. J., and Gerlach, T. M., 1994, Magmatic vapor source for sulfur dioxide released during volcanic eruptions: Evidence from Mount Pinatubo: Science, v. 265, p. 497–499.

Walraven, F., Armstrong, R. A., and Kruger, F. J., 1990, A chronostratigraphic framework for the north-central Kaapvaal craton, the Bushveld complex and the Vredefort structure: Tectonophysics, v. 171, p. 23–48.

Walter, M. R., 1980, Adelaidean and Early Cambrian stratigraphy of the southwestern Georgina Basins: correlation chart and explanatory notes: Australian Bureau of Mineral Resources Geology and Geophysics Report 214, 21 p.

Walter, M. R., Veevers, J. J., Calver, C. R., and Grey, K., 1995, Neoproterozoic stratigraphy of the Centralian Superbasin, Australia: Precambrian Research, v. 73, p. 173–195.

Wang, K., 1992, Glassy microspherulites (microtektites) from an Upper Devonian limestone: Science, v. 256, p. 1547–1550.

Wang, K., Orth, C. J., Attrep, M., Jr., Chatterton, B. D. E., Hou, H., and Geldsetzer, H. H. J., 1991, Geochemical evidence for a catastrophic biotic event at the Frasnian/Famennian boundary in south China: Geology, v. 19, p. 776–779.

Wang, K., Geldsetzer, H. H. J., and Krouse, H. R., 1994, Permian-Triassic extinction: Organic $\delta^{13}C$ evidence from British Columbia, Canada: Geology, v. 22, p. 580–584.

Wang, K., Geldsetzer, H. H. J., Goodfellow, W. D., and Krouse, H. R., 1996, Carbon and sulfur isotope anomalies across the Frasnian-Famennian extinction boundary, Alberta, Canada: Geology, v. 24, p. 187–191.

Wang Yuelin, Lu Songnian, Gao Zhenjia, Lin Weixing, and Ma Guogan, 1981, Sinian tillites of China, in Hambrey, M. J., and Harland, W. B., eds., Earth's pre-Pleistocene glacial record: Cambridge, United Kingdom, Cambridge University Press, p. 386–401.

Wanless, H. R., and Cannon, J. R., 1966, Late Paleozoic glaciation: Earth-Science Reviews, v. 1, p. 247–286.

Wanless, H. R., and Shepard, F. P., 1936, Sea level and climatic changes related to Late Paleozoic cycles: Geological Society of America Bulletin, v. 47, p. 1177–1206.

Warme, J. E., Douglas, R. G., and Winterer, E. L., eds., 1981, The Deep Sea Drilling Project: a decade of progress: Tulsa, Oklahoma, Society of Economic Paleontologists and Mineralogists Special Publication No. 32, 564 p.

Waterhouse, J. G., and Flood, P. G., 1981, Poorly sorted conglomerates, breccias and diamictites in Late Palaeozoic, Mesozoic and Tertiary sediments of New Zealand, in Hambrey, M. J., and Harland, W. B., eds., Earth's glacial record: Cambridge, United Kingdom, Cambridge University Press, p. 438–446.

Weart, S. R., 1997, The discovery of the risk of global warming: Physics Today, v. 50, p. 34–40.

Webb, A. W., Coats, R. P., Fanning, C. M., and Flint, R. B., 1983, Geochronological framework of the Adelaide geosyncline, in Adelaide geosyncline sedimentary environments and tectonic settings, symposium: Geological Society of Australia, Abstracts, v. 109, p. 7–9.

Wegener, A., 1912, Die Entstehung der Kontinente: Geologische Rundschau, v. 3, p. 276–292.

Wegener, A., 1929, The origin of continents and oceans (English translation by J. Biram, 1996, of Die Entstehung der Kontinente und Ozeane, revised 4th edition, Braunschweig, Germany Friedr. Vieweg & Sohn: New York, Dover Publications, Inc., 246 p.)

Weltje, G., and de Boer, P. L., 1993, Astronomically induced paleoclimate oscillations reflected in Pliocene turbidite deposits on Corfun (Greece): Implications for the interpretation of higher order cyclicity in ancient turbidite systems: Geology, v. 21, p. 307–310.

White, M. E., 1990, Plant life between two ice ages down under: American Scientist, v. 78, p. 253–262.

Whittlesey, C., 1848, Note upon the drift and alluvium of Ohio and the West: American Journal of Science, v. 5, p. 205–217.

Wickham, S. M., and Peters, M. T., 1993, High $\delta^{13}C$ Neoproterozoic carbonate rocks in western North America: Geology, v. 21, p. 165–168.

Wiebols, J. H., 1955, A suggested glacial origin for the Witwatersrand conglomerates: Geological Society of South Africa Transactions, v. 58, p. 367–387.

Wignall, P. B., and Twitchett, R. J., 1996, Oceanic anoxia and the end Permian mass extinction: Science, v. 272, p. 1155–1158.

Wilde, P., and Berry, W. B. N., 1984, Destabilisation of the oceanic density structure and its significance to marine "extinction" events: Palaeogeography, Palaeoclimatology, Palaeoecology, v. 48, p. 143–162.

Wilde, P., and Berry, W. B. N., 1988, Sulfur-isotope anomaly associated with the Frasnian-Famennian extinction, Medicine Lake, Alberta, Canada: Comment: Geology, v. 16, p. 86–88.

Wilgus, C. K., Hastings, B. S., Kendall, C. G. St. C., Posamentier, H. W., Ross, C. A., and Van Wagoner, J. C., eds., 1988, Sea-level changes: An integrated approach: Tulsa, Oklahoma, Society of Economic Paleontologists and Mineralogists Special Publication No. 42, 407 p.

Wilkinson, B. H., and Walker, J. C. G., 1989, Phanerozoic cycling of sedimentary carbonate: American Journal of Science, v. 289, p. 525–548.

Williams, D. Michael, 1980, Evidence for glaciation in the Ordovician rocks of western Ireland: Geological Magazine, v. 117, p. 81–86.

Williams, Darren M., Kasting, J. F., and Frakes, L. A., 1998, Low-latitude glaciation and rapid changes in the Earth's obliquity explained by obliquity-oblateness feedback: Nature, v. 396, p. 453–455.

Williams, G. E., 1975a, Possible relation between periodic glaciation and the flexure of the galaxy: Earth and Planetary Science Letters, v. 26, p. 361–369.

Williams, G. E., 1975b, Late Precambrian glacial climate and the Earth's obliquity: Geological Magazine, v. 112, p. 441–444.

Williams, G. E., 1979, Sedimentology, stable-isotope geochemistry and palaeoenvironment of dolostones capping late Precambrian glacial sequences in Australia: Geological Society of Australia Journal, v. 26, p. 377–386.

Williams, G. E., 1986a, Precambrian permafrost horizons as indicators of palaeoclimate: Precambrian Research, v. 32, p. 233–242.

Williams, G. E., 1986b, The Acraman impact structures: Source of ejecta in Late Precambrian shales, South Australia: Science, v. 233, p. 200–203.

Williams, G. E., 1988, Cyclicity in the late Precambrian Elatina Formation, South Australia: Solar or tidal signature?: Climatic Change, v. 13, p. 117–128.

Williams, G. E., 1989, Late Precambrian tidal rhythmites in South Australia and the history of the Earth's rotation: Geological Society of London Journal, v. 146, p. 97–111.

Williams, G. E., 1990, Tidal rhythmites: Key to the history of the Earth's rotation and lunar orbit: Journal of Physics of Earth, v. 38, p. 475–491.

Williams, G. E., 1993, History of the Earth's obliquity: Earth-Science Reviews, v. 34, p. 1–45.

Williams, G. E., 1994, The enigmatic Late Proterozoic glacial climate: an Australian perspective. in Deynoux, M., Miller, J. M. G., Domack, E. W.,

Eyles, N., Fairchild, I. J., and Young, G. M., eds., Earth's glacial record: Cambridge, United Kingdom, Cambridge University Press, p. 146–164.

Williams, G. E., and Schmidt, P. W., 1997a, Paleomagnetism of the Paleoproterozoic Gowganda and Lorrain formations, Ontario: Low paleolatitude for Huronian glaciation: Earth and Planetary Science Letters, v. 153, p. 157–169.

Williams, G. E., and Schmidt, P. W., 1997b, Palaeomagnetic dating of the sub-Torridon Group weathering profiles, NW Scotland: Verification of Neoproterozoic paleosols: Geological Society of London Journal, v. 154, p. 987–997.

Williams, G. E., and Sonett, C. P., 1985, Solar signature in sedimentary cycles from the late Precambrian Elatina Formation, Australia: Nature, v. 318, p. 523–527.

Williams, G. E., and Tonkin, D. G., 1985, Periglacial structures and palaeoclimatic significance of a late Precambrian block field in the Cattle Grid copper mine, Mount Gunson, South Australia: Australian Journal of Earth Sciences, v. 32, p. 287–300.

Williams, G. E., Schmidt, P. W., and Embleton, B. J. J., 1995, Comment on 'The Neoproterozoic (1100–540 Ma) glacial intervals: No more snowball Earth?' by Meert, J. G., and van der Voo, R., 1995: Earth and Planetary Science Letters, v. 131, p. 115–122.

Williams, I. R., 1992, Geology of the Savory basin, Western Australia: Geological Survey of Western Australia Bulletin, v. 141, 115 p.

Wilson, C. B., and Harland, W. B., 1964, The Polarisbreen Series and other evidences of late Pre-Cambrian ice ages in Spitzbergen: Geological Magazine, v. 101, p. 198–219.

Windley, B. F., 1984, The evolving continents: London, John Wiley and Sons, 2nd edition, 399 p.

Windley, B. F., 1993, Uniformitarianism today: Plate tectonics is the key to the past: Geological Society of London Journal, v. 150, p. 7–19.

Winterer, E. L., 1964, Late Precambrian pebbly mudstone in Normandy, France: Tillite or tilloid?, *in* Nairn, A. E. M., ed., Problems of Palaeoclimatology: London, Interscience, p. 159–178, 185–187.

Wise, D. U., 1994, Neoproterozoic crustal thinning, emergence of continents, and origin of the Phanerozoic ecosystem: A model: Comment: Geology, v. 22, p. 87–88.

Woodcock, N. H., and Smallwood, S. D., 1987, Late Ordovician shallow marine environments due to glacio-eustatic regression: Scrach Formation, Mid-Wales: Geological Society of London Journal, v. 144, p. 393–400.

Wopfner, H., and Casshyap, S. M., 1997, Transition from freezing to subtropical climates in the Permo-Carboniferous of Afro-Arabia and India, *in* Martini, I. P., ed., Late glacial and postglacial environmental changes: New York, Oxford University Press, p. 192–212.

Wopfner, H., and Kreuser, T., 1986, Evidence for late Palaeozoic glaciation in southern Tanzania: Palaeogeography, Palaeoclimatology, Palaeoecology, v. 56, p. 259–275.

Worku, T., and Astin, T. R., 1992, The Karoo sediments (Late Palaeozoic to Early Jurassic) of the Ogaden Basin, Ethiopia: Sedimentary Geology, v. 76, p. 7–21.

Worsley, T. R., and Kidder, D. L., 1991, First-order coupling of paleogeography and CO_2, with global surface temperature and its latitudinal contrast: Geology, v. 19, p. 1161–1164.

Worsley, T. R., and Nance, R. D., 1989, Carbon redox and climate control through Earth history: A speculative reconstruction: Palaeogeography, Palaeoclimatology, Palaeoecology, v. 75, p. 259–282.

Worsley, T. R., Moody, J. B., and Nance, R. D., 1985, Proterozoic to Recent tectonic tuning of biogeochemical cycles, *in* Sundquist, E. T., and Broecker, W. S., eds., The carbon cycle and atmospheric CO_2: Natural variations Archean to present: Washington, D.C., American Geophysical Union, Geophysical Monograph 32, p. 561–572.

Worsley, T. R., Nance, R. D., and Moody, J. B. 1986, Tectonic cycles and the history of the Earth's biogeochemical and paleoceanographic record: Paleoceanography, v. 1, p. 233–263.

Wright, H. E., Jr., 1989, The Quaternary, *in* Bally, A. W., and Palmer, A. R., eds., The Geology of North America—An overview: Boulder, Colorado,

Geological Society of America, The Geology of North America, v. A, p. 513–536.

Wright, L. A., and Prave, A. R., 1993, Proterozoic–Early Cambrian tectonostratigraphic record in the Death Valley Region, California-Nevada, *in* Reed, J. C., Jr., Bickford, M. E., Houston, R. S., Link, P. K., Rankin, D. W., Sims, P. K., and Van Schmus, W. R., eds., Precambrian: Conterminous U.S.: Boulder, Colorado, Geological Society of America, The Geology of North America, v. C-2, p. 529–533.

Wyatt. A. R., 1995, Late Ordovician extinctions and sea-level change: Geological Society of London Journal, v. 152, p. 899–902.

Xiao, S., Zhang, Y., and Knoll, A. H., 1998, Three-dimensional preservation of algae and animal embryos in a Neoproterozoic phosphorite: Nature, v. 391, p. 553–558.

Yapp, C. J., and Poths, H., 1992, Ancient atmospheric CO_2 pressures inferred from natural goethites: Nature, v. 355, p. 342–344.

Yeo, G. M., 1981, The Late Proterozoic Rapitan glaciation in the northern Cordillera, *in* Campbell, F. H. A., ed., Proterozoic basins of Canada: Geological Survey of Canada Paper 81-10, p. 25–46.

Yeo, G. M., 1986, Iron-formation in the late Proterozoic Rapitan Group, Yukon and Northwest Territories, *in* Morin, J. A., ed., Mineral deposits of the northern Cordillera: Montreal, Quebec, Canadian Institute of Mining and Metallurgy Special Volume 37, p. 142–153.

Youle, J. C., Watney, W. L., and Lambert, L. L., 1994, Stratal hierarchy and sequence stratigraphy—Middle Pennsylvanian, southwestern Kansas, U.S.A., *in* Klein, G. D., ed., Pangea: Paleoclimate, tectonics, and sedimentation during accretion, zenith, and breakup of a supercontinent: Geological Society of America Special Paper 288, p. 267–286.

Young, G. M., 1968, Sedimentary structures in Huronian rocks of Ontario: Palaeogeography, Palaeoclimatology, Palaeoecology, v. 4, p. 125–153.

Young, G. M., 1969, Geochemistry of Early Proterozoic tillites and argillites of the Gowganda Formation, Ontario, Canada: Geochimica et Cosmochimica Acta, v. 33, p. 483–492.

Young, G. M., 1970, An extensive early Proterozoic glaciation in North America?: Palaeogeography, Palaeoclimatology, Palaeoecology, v. 7, p. 85–101.

Young, G. M., ed., 1973, Huronian stratigraphy and sedimentation: Montreal, Quebec, Geological Association of Canada Special Paper 12, 271 p.

Young, G. M., 1976a, Discussion of "Late Precambrian mixtites: glacial and/or non-glacial?" by Schermerhorn, L. J. G.: American Journal of Science, v. 276, p. 366–370.

Young, G. M., 1976b, Iron-formation and glaciogenic rocks of the Rapitan Group, Northwest Territories, Canada: Precambrian Research, v. 3, p. 137–158.

Young, G. M., 1981a, The Early Proterozoic Gowganda Formation, Ontario, Canada, *in* Hambrey, M. J., and Harland, W. B., eds., Earth's pre-Pleistocene glacial record: Cambridge, United Kingdom, Cambridge University Press, p. 807–812.

Young, G. M., 1981b, Diamictites of the Early Proterozoic Ramsay Lake and Bruce Formations, north shore of Lake Huron, Ontario, Canada, *in* Hambrey, M. J., and Harland, W. B., eds., Earth's pre-Pleistocene glacial record: Cambridge, United Kingdom, Cambridge University Press, p. 813–816.

Young, G. M., 1982, The Late Proterozoic Tindir Group, east central Alaska; evolution of a continental margin: Geological Society of America Bulletin, v. 93, p. 759–783.

Young, G. M., 1983, Tectono-sedimentary history of Early Proterozoic rocks of the northern Great Lakes region, *in* Medaris, L. G., Jr., ed., Early Proterozoic geology of the Great Lakes Region: Geological Society of America Memoir 160, p. 15–34.

Young, G. M., 1984, Proterozoic plate tectonics in Canada with emphasis on evidence for a Late Proterozoic rifting event: Precambrian Research, v. 25, p. 233–256.

Young, G. M., 1988, Proterozoic plate tectonics, glaciation and iron-formations: Sedimentary Geology, v. 58, p. 127–144.

Young, G. M., 1991, The geologic record of glaciation: Relevance to the climatic history of Earth: Geoscience Canada, v. 18, p. 100–108.

Young, G. M., 1992a, Late Proterozoic stratigraphy and the Canada-Australia connection: Geology, v. 20, p. 215–218.

Young, G. M., 1992b, Neoproterozoic glaciation in the Broken Hill area, New South Wales, Australia: Geological Society of America Bulletin, v. 104, p. 840–850.

Young, G. M., 1995a, Are Neoproterozoic glacial deposits preserved on the margins of Laurentia related to the fragmentation of two supercontinents?: Geology, v. 23, p. 153–156.

Young, G. M., 1995b, Are Neoproterozoic glacial deposits preserved on the margins of Laurentia related to the fragmentation of two supercontinents?: Reply: Geology, v. 23, p. 960–961.

Young, G. M., and Gostin, V. A., 1988, Stratigraphy and sedimentology of Sturtian glacigenic deposits in the western part of the North Flinders basin, South Australia: Precambrian Research, v. 39, p. 151–170.

Young, G. M., and Gostin, V. A., 1989, An exceptionally thick upper Proterozoic (Sturtian) glacial succession in the Mount Painter area, South Australia: Geological Society of America Bulletin, v. 101, p 834–845.

Young, G. M., and Gostin, V. A., 1990, Sturtian glacial deposition in the vicinity of the Yankaninna anticline, North Flinders Basin, South Australia: Australian Journal of Earth Sciences, v. 37, p. 447–458.

Young, G. M., and Gostin, V. A., 1991, Late Proterozoic (Sturtian) succession of the North Flinders Basin, South Australia; An example of temperate glaciation in an active rift setting, *in* Anderson, J. B., and Ashley, G. M., eds., Glacial marine sedimentation; Paleoclimatic significance: Geological Society of America Special Paper 261, p. 207–222.

Young, G. M., and McLennan, S. M., 1981, Early Proterozoic Padlei Formation, Northwest Territories, Canada, *in* Hambrey, M. J., and Harland, W. B., eds., Earth's pre-Pleistocene glacial record: Cambridge, United Kingdom, Cambridge University Press, p. 790–794.

Young, G. M., and Nesbitt, H. W., 1985, The Gowganda Formation in the southern part of the Huronian outcrop belt, Ontario, Canada: Stratigraphy, depositional environments and regional tectonic significance: Precambrian Research, v. 29, p. 265–301.

Young, G. M., and Nesbitt, H. W., 1999, Paleoclimatology and provenance of the glaciogenic Gowganda Formation (Paleoproterozoic), Ontario, Canada: A chemostratigraphic approach: Geological Society of America Bulletin, v. 111, p. 264–274.

Young, G. M., Jefferson, C. W., Delaney, G. D., and Yeo, G. M., 1979, Middle and Late Proterozoic evolution of the northern Canadian Cordillera and shield: Geology, v. 7, p. 125–128.

Young, G. M., von Brunn, V., Gold, D. J. C., and Minter, W. E. L., 1998, Earth's oldest reported glaciation: Physical and chemical evidence from the Archean Mozaan Group (~2.9 Ga) of South Africa: Journal of Geology, v. 102, p. 523–538.

Yugan, J., Wardlaw, B. R., Glenister, B. F., and Kotlyar, G. V., 1997, Permian stratigraphic subdivisions: Episodes, v. 20, p. 10–15.

Zachos, J. C., Flower, B. P., and Paul, H., 1997, Orbitally paced climate oscillations across the Oligocene/Miocene boundary: Nature, v. 388, p. 567–570.

Zhang, H., and Zhang, W., 1985, Palaeomagnetic data, late Precambrian magneto-stratigraphy and tectonic evolution of eastern China: Precambrian Research, v. 29, p. 65–75.

Zheng Zhaochang, Li Yuzhen, Lu Songnian, and Li Huaikun, 1994, Lithology, sedimentology and genesis of the Zhengmuguan Formation of Ningxia, China, *in* Deyoux, M., Miller, J. M. G., Domack, E. W., Eyles, N., Fairchild, I. J., and Young, G. M., eds.: Cambridge, United Kingdom, Cambridge University Press, p. 101–108.

Ziegler, A. M., 1990, Phytogeographic patterns and continental configurations during the Permian Period, *in* McKerrow, W. S., and Scotese, C. R., eds., Palaeozoic palaeogeography and biogeography: Geological Society of London Memoir 12, p. 363–382.

Ziegler, A. M., Hulver, M. L., and Rowley, D. G., 1997, Permian world topography and climate, *in* Martini, I. P., ed., Late Glacial and Postglacial Environmental Changes: New York, Oxford University Press, p. 111–146.

Ziegler, P. A., 1959, Frühpaläozoische Tillite im östlichen Yukon-Territorium (Kanada): Eclogae Geologicae Helveticae, v. 52, p. 735–741.

MANUSCRIPT ACCEPTED BY THE SOCIETY MARCH 2, 1999